Aktuelle Forschung Medizintechnik

Editor-in-Chief:
Th. M. Buzug, Lübeck, Deutschland

Unter den Zukunftstechnologien mit hohem Innovationspotenzial ist die Medizintechnik in Wissenschaft und Wirtschaft hervorragend aufgestellt, erzielt überdurchschnittliche Wachstumsraten und gilt als krisensichere Branche. Wesentliche Trends der Medizintechnik sind die Computerisierung, Miniaturisierung und Molekularisierung. Die Computerisierung stellt beispielsweise die Grundlage für die medizinische Bildgebung, Bildverarbeitung und bildgeführte Chirurgie dar. Die Miniaturisierung spielt bei intelligenten Implantaten, der minimalinvasiven Chirurgie, aber auch bei der Entwicklung von neuen nanostrukturierten Materialien eine wichtige Rolle in der Medizin. Die Molekularisierung ist unter anderem in der regenerativen Medizin, aber auch im Rahmen der sogenannten molekularen Bildgebung ein entscheidender Aspekt. Disziplinen übergreifend sind daher Querschnittstechnologien wie die Nano- und Mikrosystemtechnik, optische Technologien und Softwaresysteme von großem Interesse.

Diese Schriftenreihe für herausragende Dissertationen und Habilitationsschriften aus dem Themengebiet Medizintechnik spannt den Bogen vom Klinikingenieurwesen und der Medizinischen Informatik bis hin zur Medizinischen Physik, Biomedizintechnik und Medizinischen Ingenieurwissenschaft.

Editor-in-Chief:
Prof. Dr. Thorsten M. Buzug
Institut für Medizintechnik,
Universität zu Lübeck

Editorial Board:
Prof. Dr. Olaf Dössel
Institut für Biomedizinische Technik,
Karlsruhe Institute for Technology

Prof. Dr. Heinz Handels
Institut für Medizinische Informatik,
Universität zu Lübeck

Prof. Dr.-Ing. Joachim Hornegger
Lehrstuhl für Mustererkennung,
Universität Erlangen-Nürnberg

Prof. Dr. Marc Kachelrieß
German cancer Research Center,
Heidelberg

Prof. Dr. Edmund Koch,
Klinisches Sensoring und Monitoring,
TU Dresden

Prof. Dr.-Ing. Tim C. Lüth
Micro Technology
and Medical Device Technology,
TU München

Prof. Dr. Dietrich Paulus
Institut für Computervisualistik,
Universität Koblenz-Landau

Prof. Dr. Bernhard Preim
Institut für Simulation und Graphik,
Universität Magdeburg

Prof. Dr.-Ing. Georg Schmitz
Lehrstuhl für Medizintechnik,
Universität Bochum

Marlitt Erbe

Field Free Line Magnetic Particle Imaging

 Springer Vieweg

Marlitt Erbe
University of Lübeck
Germany

Dissertation University of Lübeck, 2013

ISBN 978-3-658-05336-9   ISBN 978-3-658-05337-6 (eBook)
DOI 10.1007/978-3-658-05337-6

The Deutsche Nationalbibliothek lists this publication in the Deutsche Nationalbibliografie;
detailed bibliographic data are available in the Internet at http://dnb.d-nb.de.

Library of Congress Control Number: 2014933589

Springer Vieweg
© Springer Fachmedien Wiesbaden 2014

Printed on acid-free paper

Springer Vieweg is a brand of Springer DE.
Springer DE is part of Springer Science+Business Media.
www.springer-vieweg.de

Preface by the Series Editor

The book *Field Free Line Magnetic Particle Imaging* by Dr. Marlitt Erbe is the 8th volume of the new Springer-Vieweg series of excellent theses in medical engineering. The thesis of Dr. Erbe has been selected by an editorial board of highly recognized scientists working in that field.

The Springer-Vieweg series aims to establish a forum for Monographs and Proceedings on Medical Engineering. The series publishes works that give insights into the novel developments in that field. Prospective authors may contact the Series Editor about future publications within the series at:

Prof. Dr. Thorsten M. Buzug
Series Editor Medical Engineering
Institute of Medical Engineering
University of Lübeck
Ratzeburger Allee 160
23562 Lübeck
Web: www.imt.uni-luebeck.de
Email: buzug@imt.uni-luebeck.de

Geleitwort

Magnetic-Particle-Imaging ist eine bildgebende Methode, die den Tracer-Verfahren zuzuordnen ist. Eine Suspension von Dextran-umhüllten Eisenoxid-basierten Nanopartikeln soll dabei dem Organismus appliziert werden. Die hoch sensitive und echtzeitfähige Abbildung der räumlichen Verteilung der Partikel ist das Ziel der Methode. Dazu wird über eine Maxwellspulenpaar-anordnung zunächst ein Selektionsfeld erzeugt. Der Null-Durchgang des Selektionsfeldgradienten wird dann über ein sogenanntes Drive-Field periodisch im Raum verschoben, so dass sich eine gewünschte Abtasttrajektorie ergibt, die der Spur dieses Null-Durchganges (dem feldfreien Punkt - FFP) entspricht. Die Änderung der Magnetisierung der Nanopartikel wird über eine Empfangsspuleneinrichtung detektiert. Dabei macht die Nichtlinearität der Magnetisierung, die über die Langevin-Theorie in erster Näherung gut beschrieben werden kann, die Messung der Magnetisierungsänderung erst möglich, denn sie erzeugt eine Reihe von Oberwellen, die vom anregenden Drive-Field nicht überlagert werden. Da nicht nur die Magnetisierungsänderung der Partikel im direkten FFP-Durchgang, sondern auch die der Partikel etwas außerhalb zum Empfangssignal beitragen, ist eine Entfaltung des Signals erforderlich. Während das Prinzip bis hierher einfach zu beschreiben und zu verstehen ist, liegt die Kunst der Bildgebung im Detail. Auch davon berichtet das vorliegende Buch von Marlitt Erbe. Sie beschreibt die Prinzipien dieser neuen Modalität sehr präzise, diskutiert eine alternative Signalkodierung, bei der die Spur des Null-Durchgangs des Selektionsfeldgradienten nicht einem Punkt, sondern einer feldfreien Linie (FFL) entspricht, und stellt viele Originalbeiträge vor. Gleichzeitig gibt dieses Werk auch eine detaillierte Einführung in das Forschungsgebiet MPI und in die FFL-Bildgebung, die eine deutliche Sensitivitätssteigerung im Vergleich zu der FFP-Bildgebung verspricht und die Möglichkeit eröffnet, effiziente Radon-basierte Rekonstruktionsalgorithmen zu nutzen. Frau Erbe beschreibt die Innovationen im Bereich des FFL-Scanner-Designs und gibt eine Einführung in die unterschiedlichen FFL-Bildgebungstechniken. Das vorliegende Werk ist als herausragend zu beurteilen. Sprachlich schnörkellos reihen sich Originalbeiträge in dieser Arbeit aneinander.

Prof. Dr. Thorsten M. Buzug

Acknowledgements

To my mentor Thorsten Buzug for his ability to create an environment for me to prosper in scientific, professional, and personal regard.

To my colleagues and dear friends Timo Sattel and Tobias Knopp for their constant support and the most delightful collaboration.

To the most dedicated and bright students Matthias Weber and Klaas Bente.

To Bärbel Kratz and Andreas Mang for their enduring interpersonal support.

To the MPI team at our institute for their ideas and dedication.

To my parents Helga and Peter Erbe for keeping me from concerns.

To Anni Erbe for being a timeless role model of kindness.

To my dear friends Annika and Christina Pagel, Sophia Beuster, and Eva Dahlke for keeping me open-minded.

To my brother Janni Erbe for his wits and for proof-reading.

To my beloved Leroy Jönsson.

Abstract

Modern medicine is intrinsically tied to technology. Not only medical therapy, but also medical diagnostics considerably benefit from technological progress. This connection is impressively demonstrated by the development of medical imaging technologies, which have revolutionized a countless number of medical applications. Since the very first x-ray images were produced in 1895 [R98], it was not until the 1960s and 1970s that the powerful medical imaging technologies computed tomography (CT) and magnetic resonance imaging (MRI) were introduced [Lau73, Hou73, Cor63, Cor64]. Since those first steps, groundbreaking progress in spatial and temporal resolution as well as dose issues have been achieved. CT offers slice images of the human body with the use of x-ray radiation [Buz08]. While in CT a high spatial and temporal resolution is achieved, the disadvantage of this method is the exposure of the patient and the physician to ionizing radiation. MRI in contrast, works without the need of ionizing radiation, but suffers from poor temporal resolution.

In 2005, B. Gleich and J. Weizenecker introduced a novel medical imaging technology called magnetic particle imaging (MPI) [GW05], which combines high spatial and temporal resolution with abdication of harmful radiation. MPI images the distribution of a superparamagnetic iron oxide (SPIO) nanoparticle tracer material within a patient. It intrinsically provides three dimensional (3D) data with a sub-millimeter resolution and is capable of performing real-time imaging. In the very first *in-vivo* MPI experiments, 3D images of a beating mouse heart were produced [WGR+09] demonstrating the potential held by this fascinating new tomographic method to considerably improve imaging of fast blood-flow phenomena.

As originally introduced, MPI applies a sensitive spot encoding scheme using the SPIO particle signal emanating from a field free point (FFP). To enhance sensitivity, however, signal acquisition along a field free line (FFL) was introduced in 2008 and the expected gain in sensitivity was demonstrated in a simulation study [WGB08]. Unfortunately, the setup of electromagnetic coils proposed in [WGB08] was highly inefficient and practical implementation of an FFL imaging system was therefore doubted even by its inventors.

After several scanner optimizations [KBS+10a, KEB+10], however, an efficient design generating the magnetic fields of a complete FFL trajectory was implemented, tested and evaluated as part of this thesis for the first time [EKS+11a]. The results demonstrate that efficient FFL generation is feasible with respect to practical implementation and motivated ongoing research in the field of dynamic FFL imaging.

The presented thesis provides a detailed introduction into the young research field of MPI and FFL imaging in particular. After describing the innovations concerning FFL scanner design and introducing FFL imaging techniques, a mathematical description on magnetic field generation for FFL imaging in MPI is derived. From this, the efficient scanner design introduced in [KEB+10] and used for the implementation of the first FFL field demonstrator [EKS+11a] is motivated from a purely mathematical point of view. To substantiate the simulation studies on magnetic FFL generation with a proof-of-concept, the FFL field demonstrator is presented, which provides the world's first experimentally generated rotated and translated magnetic FFL field complying with the requirements for FFL reconstruction.

Towards the implementation of a dynamic FFL imaging system, a scanner design of enhanced quality and efficiency is proposed in this thesis [ESK+11]. Regarding implementation of an assembly of electromagnetic coils and permanent magnets constituting the basis of an FFL imaging system, the quality of the magnetic fields as well as the overall electrical power consumption are of great importance. In a detailed analysis of both of these parameters, curved rectangular coils are identified to provide an ideal combination of high magnetic field quality and low electrical power consumption [ESKB12]. In fact, with the design proposed in this thesis, the magnetic field quality is enhanced by a factor of almost five, while simultaneously electrical power consumption is reduced by a factor of almost four compared to the conventional design consisting of circular coils. Since manufacturing of curved rectangular coils is more challenging than manufacturing of circular coils, a test assembly is presented, which demonstrates feasibility of coil production and proofs that no additional field errors are introduced due to manufacturing inaccuracy. Based on these results a design decision for the installation of the first dynamic FFL mouse scanner is presented.

Magnetic field quality optimization is of special importance regarding efficient Radon-based reconstruction methods, which arise for a line detection scheme [KES+11] and demand for high gradient homogeneity. A mathematical derivation of FFL-data transformation into Radon space and a simulation study on the influence of magnetic field quality on the achieved image quality is presented. It is shown that magnetic field quality optimization leads to a reduction of artifact formation in Radon-based FFL reconstruction [EKSB12].

Based on the design optimization presented in this thesis, the installation of the first dynamic FFL scanner is realized. The complete MPI signal chain is installed and presented. Due to the design optimization, a higher magnetic field gradient is generated at reduced electrical power consumption compared to the results of [EKS+11a] at enhanced magnetic field quality [EWSB13]. After initiation of all components, dynamic FFL mouse imaging will be realized with the presented scanner setup.

This thesis constitutes a first step towards the establishment of dynamic FFL imaging in MPI. The results do not only give rise to optimism concerning feasibility of magnetic FFL generation, which has been doubted during the invention of the FFL encoding scheme, but also reveal various research aspects holding great potential to improve the results achieved using FFL imaging in MPI such as scanner optimization and efficient Radon-based reconstruction.

Kurzfassung

Die moderne Medizin ist untrennbar mit dem technologischen Fortschritt verbunden. Nicht nur die medizinische Therapie, sondern auch die medizinische Diagnostik profitieren wesentlich von neuen technologischen Entwicklungen. Diese Verbindung wird besonders deutlich durch die Entwicklung medizinischer Bildgebungsverfahren demonstriert, die eine hohe Anzahl an medizinischen Anwendungen revolutioniert haben. Obwohl die ersten Röntgenbilder im Jahr 1895 erzeugt wurden [R98], dauerte es bis in die sechziger und siebziger Jahre des 20. Jahrhunderts, bis die leistungsstarken medizinischen Bildgebungsverfahren, die Computertomographie (CT) und die Magnetresonanztomographie (MRT), vorgestellt wurden [Lau73, Hou73, Cor63, Cor64]. Seit diesen ersten Schritten wurden bahnbrechende Fortschritte im Bereich der räumlichen und zeitlichen Auflösung sowie in Fragen der Dosis erzielt. In der CT werden mit Hilfe von Röntgenstrahlung Schichtbilder des menschlichen Körpers erzeugt [Buz08]. Während CT eine hohe räumliche und zeitliche Auflösung erzielt, ist der Nachteil dieser Methode, dass sowohl der Patient als auch der behandelnde Arzt einer nicht vernachlässigbaren Strahlenbelastung ausgesetzt sind. MRT dagegen kommt ohne Strahlenbelastung aus, bietet jedoch nur eine geringe zeitliche Auflösung.

Im Jahr 2005 stellten B. Gleich und J. Weizenecker das neuartige, medizinische Bildgebungsverfahren Magnetic-Particle-Imaging (MPI) vor [GW05], das eine hohe räumliche und zeitliche Auflösung und einen Verzicht auf schädliche Strahlung verbindet. MPI bildet die Verteilung eines superparamagnetischen Eisenoxid-Tracermaterials ab, das in den Blutkreislauf eines Patienten injiziert wird. Die mit MPI erzeugten Daten enthalten drei-dimensionale (3D) Informationen über die Verteilung des Tracermaterials innerhalb des menschlichen Körpers mit einer räumlichen Auflösung unterhalb eines Millimeters. Ein weiterer entscheidender Vorteil von MPI gegenüber etablierten Bildgebungsmodalitäten ist die Möglichkeit zur Echtzeitbildgebung. Schon in den ersten *in-vivo* MPI-Experimenten wurden Bilder eines schlagenden Mäuseherzens erzeugt [WGR+09], die das Potential dieser faszinierenden, neuen Bildgebungsmethode eindrucksvoll demonstrieren, die Visualisierung von schnellen Blutflussphänomenen wesentlich zu verbessern.

MPI, wie ursprünglich eingeführt, verwendet das Partikelsignal ausgehend von einem feldfreien Punkt (FFP) [GW05]. Um die Sensitivität zu steigern, wurde jedoch im Jahr 2008 die Signalaufnahme entlang einer feldfreien Linie (FFL) vorgeschlagen und die erwartete Sensitivitätssteigerung in einer Simulationsstudie demonstriert [WGB08]. Unglücklicherweise war der Aufbau aus elektromagnetischen Spulen, der in [WGB08] verwendet wurde, höchst ineffizient und die Machbarkeit der Implementierung eines FFL-Bildgebungssystems wurde daher sogar von seinen Erfindern angezweifelt.

Nach mehreren Optimierungsschritten in Bezug auf das FFL-Scannerdesign [KBS+10a, KEB+10] wurde schließlich als Teil dieser Arbeit der Experimentalaufbau eines effizienten Spulendesigns realisiert, getestet und validiert, mittels dessen erstmals die magnetischen Felder einer kompletten FFL-Trajektorie generiert werden konnten [EKS+11a]. Die erzielten Resultate stellen einen Machbarkeitsnachweis für die effiziente FFL-Magnetfelderzeugung dar und motivieren weitere Forschungsarbeiten im Bereich der dynamischen FFL-Bildgebung.

Diese Arbeit gibt eine detaillierte Einführung in das junge Forschungsgebiet MPI und in die FFL-Bildgebung im Speziellen. Nach einer Beschreibung der Innovationen im Bereich des FFL-Scannerdesigns und einer Einführung der unterschiedlichen FFL-Bildgebungstechniken, wird eine mathematische Beschreibung der Magnetfelderzeugung für die FFL-Bildgebung hergeleitet. Davon ausgehend wird das effiziente FFL-Scannerdesign, welches in [KEB+10] vorgestellt und für die Implementierung des ersten FFL-Felddemonstrators genutzt wird [EKS+11a], von einem rein mathematischen Gesichtspunkt aus motiviert. Um die Simulationsstudien zur FFL-Magnetfeldgenerierung mit einem Machbarkeitsnachweis zu untermauern, wird der FFL-Felddemonstrator realisiert. Mit diesem wird erstmals die Erzeugung, Rotation und Translation eines FFL-Feldes ohne mechanische Bewegung des Aufbaus realisiert, was die Voraussetzungen für eine dynamische FFL-Bildrekonstruktion darstellt.

Auf dem Weg in Richtung der Implementierung eines dynamischen FFL-Bild-gebungssystems, wird in dieser Arbeit ein Scannerdesign vorgeschlagen, das sich von seinen Vorgängern durch eine deutliche Qualitäts- und Effizienzstei-gerung abhebt [ESK+11]. Im Hinblick auf die Implementierung eines Aufbaus aus elektromagnetischen Spulen und Permanentmagneten, der die Basis für ein FFL-Bildgebungssystem bildet, sind die Magnetfeldqualität und die elek-trische Verlustleistung von entscheidender Wichtigkeit. In einer detaillierten Analyse dieser beiden Parameter wird gezeigt, dass ein Scanner bestehend aus gekrümmten Rechteckspulen eine ideale Kombination aus einer hohen Magnet-feldqualität und einer niedrigen elektrischen Verlustleistung bietet [ESKB12]. Tatsächlich wird die Magnetfeldqualität mit dem vorgeschlagenen Design an-nähernd um den Faktor fünf erhöht, während die elektrische Verlustleistung verglichen mit dem konventionellen Design aus Kreisspulen fast auf ein Vier-tel reduziert wird. Da der Herstellungsprozess von gekrümmten Rechteck-spulen anspruchsvoller ist als der von gewöhnlichen Kreisspulen, wird ein Test-aufbau realisiert, der die Machbarkeit der Produktion dieser Spulen speziell für MPI zeigt. Anhand dieses Aufbaus wird mittels Magnetfeldmessungen demonstriert, dass keine Feldabweichungen aufgrund von Herstellungsunge-nauigkeiten auftreten. Basierend auf diesen Ergebnissen wird eine Designent-scheidung für den Aufbau des ersten dynamischen FFL-Mäusescanners präsen-tiert.

Magnetfeldoptimierung ist für die FFL-Bildgebung von besonderer Bedeutung, da die Möglichkeit besteht, effiziente Radon-basierte Rekonstruktionsalgorith-men anzuwenden, die das Liniendetektionsschema ausnutzen [KES+11] und eine hohe Gradientenlinearität voraussetzen. Es wird eine mathematische Her-leitung der Transformation der FFL-Daten in den Radonraum sowie eine Simu-lationsstudie zum Einfluss der Magnetfeldqualität auf die erzielte Bildqualität präsentiert. Außerdem wird gezeigt, dass durch eine Magnetfeldoptimierung Bildartefakte reduziert werden, die bei der effizienten, Radon-basierten FFL-Rekonstruktion auftreten können [EKSB12].

Basierend auf den Designoptimierungen dieser Arbeit, wird der Aufbau des ersten dynamischen FFL-Bildgebungssystems realisiert. Die gesamte MPI-Sig-nalkette wird installiert und vorgestellt. Im Vergleich zu den Ergebnissen aus [EKS+11a] wird aufgrund der Designoptimierung ein höherer Magnetfeldgra-dient bei reduzierter elektrischer Verlustleistung und gesteigerter Magnetfeld-qualität erzielt [EWSB13]. Nach Inbetriebnahme aller Komponenten, wird der präsentierte Aufbau die Erzeugung von dynamische FFL-Mausbilder ermöglichen. Diese Arbeit stellt einen ersten Schritt auf dem Weg zur Etablierung der dynamischen FFL-Bildgebung in MPI dar. Die präsentierten Ergebnisse geben nicht nur Grund zum Optimismus in Bezug auf die Machbarkeit der Magnetfeldgenerierung für die FFL-Bildgebung in MPI, welche im Zuge der Einführung dieser Methode angezweifelt wurde, sondern zeigen auch weitere vielversprechende Forschungsgebiete wie die Optimierung des FFL-Scannerdesigns und die effiziente Radon-basierte Rekonstruktion auf, die großes Potential für die FFL-Bildgebung beinhalten.

Contents

List of Figures

List of Tables

CHAPTER 1

Introduction

The technological progress achieved in medical imaging has strongly influenced medical diagnostics and therapy. To identify a specific disease pattern and decide on an appropriate method of treatment, physicians depend on information from the inside of the human body. Therefore, research interest in medical imaging technologies has significantly increased during the last decades. To gain as many information as possible, images of high spatial and temporal resolution are desirable, but at the same time, the dose of ionizing radiation applied in some imaging modalities, as for instance in computed tomography (CT) [Buz08], needs to be kept preferably low.

A method combining high spatial and temporal resolution with the abdication of harmful radiation has recently been invented by B. Gleich and J. Weizenecker [GW05]. Magnetic particle imaging (MPI) is a novel tracer-based tomographic modality taking advantage of the non-linear magnetization behavior of superparamagnetic iron oxide (SPIO) nanoparticles to detect their spatial and temporal distribution within a patient. MPI offers a real-time temporal resolution, it intrinsically provides three dimensional (3D) information of the spatial distribution of the SPIO tracer within the human body, and it does not apply harmful, ionizing radiation. These capabilities of MPI open up a number of new applications as for instance dynamic heart imaging [WGR+09, RGWB10] or interventional MPI providing online supervision in the catheter laboratory [HRG+12, HBW+12]. Furthermore, MPI has the potential to considerably improve the sentinel lymph node biopsy (SLNB) [RBK+09, FRB+10, GSLB+12, SEB+12].

In MPI, a superposition of two different magnetic fields is used to detect the spatial and temporal distribution of SPIO tracer particles within a patient. These two magnetic fields are the constant gradient selection field and the spatially homogeneous, time-varying drive field. As introduced in 2005 by the inventors of MPI [GW05], the selection field features a field free point (FFP), at which SPIO particles emit a characteristic signal due to a change in their magnetization as a response on the sinusoidal drive field exciting the SPIO tracer particles. A map of the SPIO particle distribution is obtained by steering the FFP through the field of view (FOV). In the presented thesis, an alternative encoding scheme featuring a field free line (FFL) selection field [WGB08] is discussed. This methods holds several advantages compared to the conventional FFP encoding scheme. FFL imaging provides an increased sensitivity [WGB08] and furthermore, efficient Radon-based reconstruction algorithms arise for a line detection scheme [KES+11].

1.1 Motivation

In this work, an alternative method of data acquisition in MPI is introduced, which holds great potential of considerably increasing the method's sensitivity [WGB08]. In 2008, the inventors of MPI, J. Weizenecker and B. Gleich, introduced a concept applying a magnetic FFL replacing the conventionally used FFP [WGB08]. Instead of measuring the signal induced by particles located at a sensitive spot, as it is the case in FFP imaging, the magnetization response of SPIO particles located along a line is acquired. The simultaneous encoding scheme applied in FFL imaging provides a solution for an inherent problem generally existing in sensitive spot imaging methods. This problem concerns the direct coupling of the achieved resolution and the sensitivity. In MPI, both of these parameters depend on the gradient strength of the applied magnetic selection field. In the case of FFP imaging, a high gradient strength results in a small region, in which SPIO particles contribute to the detected signal, and hence in a high spatial resolution. Unfortunately, a small region, in which SPIO particles are detected, results in a small number of SPIO particles contributing to the induced signal. Hence, a high gradient strength leads to a high spatial resolution, but lowers the signal to noise ratio (SNR) and the sensitivity at the same time.

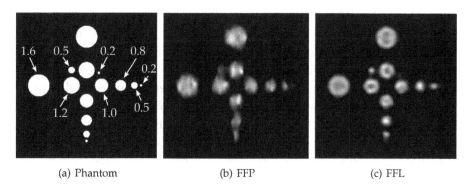

    (a) Phantom   (b) FFP   (c) FFL

Figure 1.1 MPI images simulated and reconstructed with identical parameters. (a) shows the Phantom used for simulation, where the diameter of the dots are given in mm. (b) shows the image simulated with an FFP scanner and in (c) the image simulated with an FFL scanner is illustrated. In these images, the gain in sensitivity and hence image quality using the FFL encoding scheme is demonstrated.

One solution to this problem is provided by using the magnetization response of an increased number of SPIO particles while keeping the gradient strength constant. This can be realized by applying a simultaneous encoding scheme detecting the signal induced by SPIO particles located along a line. In this FFL imaging method, the number of SPIO particles reacting to the MPI drive field is increased compared to the sensitive spot method applied in FFP imaging. A simulation study comparing FFP and FFL imaging in MPI indicates an increase in sensitivity by one order of magnitude [WGB08]. In Fig. 1.1, simulated MPI images are illustrated, which demonstrate the gain in sensitivity and image quality expected for FFL imaging compared to FFP imaging. In the underlying simulation, all parameters, i.e. scanner size and gradient strength, SPIO particle concentration and core diameter, as well as all reconstruction parameters are chosen to be equal for both, FFL and FFP imaging. Fig. 1.1(b) shows the simulated result for the FFP, while Fig. 1.1(c) is obtained applying the FFL encoding scheme. These results demonstrate the potential held by FFL imaging to increase the image quality achieved in MPI.

This thesis provides detailed theoretical considerations deriving the magnetic fields necessary for FFL imaging in MPI and an efficient FFL scanner design proposed in [KEB⁺10] is motivated from a purely mathematical point of view [EKS⁺11a]. An FFL field demonstrator is presented, which provides the first experimental proof of magnetic field generation for FFL imaging in MPI [EKS⁺11a]. Subsequently, an optimized setup of field generating electromagnetic coils is proposed, which outperforms recently introduced setups [WGB08, KSBB10, KES⁺10a, KBS⁺10a, KEB⁺10] not only concerning electrical power consumption, which is a major aspect regarding technical implementation, but also in magnetic field quality [ESKB12, EWSB13]. Furthermore, a simulation study on efficient reconstruction algorithms is provided and the effect of magnetic field quality optimization on the achieved image quality using Radon-based reconstruction algorithms is examined [EKSB12]. Finally, a device capable of dynamic FFL imaging is installed within the scope of this work, which will enable experimental validation of the dynamic FFL imaging concept for MPI.

1.2 Implemented Devices

Two hardware devices have been implemented as part of this thesis. As a first step towards the experimental proof of feasibility of MPI using the FFL encoding scheme, a field demonstrator has been constructed, realized, tested, and evaluated. This device is capable of generating a magnetic field complying with the requirements imposed by FFL reconstruction. The FFL field demonstrator provides a proof of principle for magnetic field generation for dynamic FFL imaging in MPI for the first time [EKS⁺11a] and is pictured in Fig. 1.2(a).

Furthermore, a dynamic FFL mouse scanner of the optimized design proposed in this thesis has been installed. Fig. 1.2(b) shows a photo of the center piece of the scanner, i.e. the transmit and receive coil assembly mounted inside of the scanner housing, which ensures stability of the separate components as well as appropriate cooling. Compared to the FFL field demonstrator, the FFL scanner generates an FFL field of higher gradient strength at lower electrical power consumption and improved magnetic field quality and gradient linearity [EWSB13]. Within the scope of the presented thesis, the complete MPI signal chain, i.e. transmit and receive chain, is implemented. The generated magnetic fields are measured and evaluated and show high agreement with the predictions from simulations. After simultaneous initiation of all scanner components, dynamic FFL imaging will be performed with the presented setup.

(a)   (b)

Figure 1.2 (a) First FFL field demonstrator implemented within the scope of this thesis and (b) dynamic FFL scanner implemented within the scope of this thesis an the Master's thesis of M. Weber [Web12].

1.3 Publications

Parts of this thesis have been published in scientific journals. The generation of a static magnetic FFL field with the implemented FFL field demonstrator is discussed in [KES+10b]. A mathematical description of magnetic field generation for dynamic FFL imaging in MPI as well as experimental generation of an arbitrarily rotated and translated FFL field is provided in [EKS+11a]. A general feasibility study on Radon-based FFL reconstruction is presented in [KES+11]. A simulation study on an optimized FFL imaging device consisting of curved rectangular coils is proposed in [ESB13] and the installed FFL scanner setup is implemented and evaluated within the scope of [EWSB13].

In addition, parts of this thesis have been presented on scientific conferences [EKB+10, EKS+11b, ESK+11, EGSB12, EKSB12, ESB12b, ESB12a, ESKB12]. Furthermore, scientific journal publications [KEB+10, KBS+11a, HSE+12, BBE+12], books [BBK+10, BBB+10], a book chapter [BSE+11a], and scientific conference publications [BKS+10a, BSK+10a, BSK+10b, LBBE+10, BKS+10b, KBS+10a, BSE+10, KES+10a, SKB+10, KBS+11b, SBE+11, SBE+11, BSE+11b, SEB+12, SHEB13, WEB+54] have been achieved under collaboration of the author.

1.4 Structuring

After a detailed introduction into MPI in general provided in chapter 2 and into the use of the alternative FFL encoding in particular presented in chapter 3, this thesis mainly covers five research topics

Chapter 4 Mathematical derivation of magnetic field generation for dynamic FFL imaging in MPI.

Chapter 5 Implementation of the first FFL field demonstrator providing a proof of feasibility for magnetic field generation for dynamic FFL imaging in MPI.

Chapter 6 Analysis of magnetic field quality and electrical power consumption for various coil designs leading to an FFL scanner setup outperforming recently presented designs by a factor of almost five in magnetic field quality and a factor of almost four in electrical power consumption.

Chapter 7 Simulation study on efficient Radon-based reconstruction algorithms arising for a line detection scheme. The influence of magnetic field quality on the achieved image quality is analyzed and it is shown that magnetic field quality optimization leads to image artifact reduction.

Chapter 8 Installation of a dynamic FFL mouse scanner

The presented thesis closes with a summary, a discussion and an outlook.

CHAPTER 2

Magnetic Particle Imaging

Magnetic particle imaging (MPI) is a young and very promising tomographic modality capable of imaging a distribution of biocompatible superparamagnetic iron oxide (SPIO) nanoparticles within a patient with a sub-millimeter resolution in three dimensions (3D) and real-time without the need of harmful, ionizing radiation [GW05, WGR+09]. The potential held by this revolutionary new tomographic method was already demonstrated in the first, impressive *in-vivo* MPI images of a beating mouse heart presented in [WGR+09]. No existing modality is capable of providing 3D real-time images of the inside of the human body. To classify MPI within established medical imaging modalities, the crucial properties of computed tomography (CT), magnetic resonance imaging (MRI) and positron emission tomography (PET) are compared to those of MPI in Fig. 2.1.

Due to these characteristics, MPI has the potential to revolutionize imaging of fast blood flow phenomena, as needed for example for the visualization of coronary artery diseases as well as interventional imaging in the catheter laboratory [HRG+12]. Another promising application is the localization of cancer cells when using MPI in the sentinal lymph node biopsy (SLNB) [RBK+09, FRB+10]. A closer look at such medical applications will be taken in the subsequent section 2.1.

To spatially and temporally resolve a distribution of SPIO tracer particles within a patient, the interaction between static and dynamic magnetic fields and the characteristic magnetization behavior of the tracer material is exploited. To in-

	CT	MRI	PET	MPI
Resolution	0.5 mm	1 mm	4 mm	< 1 mm
Sensitivity	low	low	high	high
Acquisition Time	1 s	10 s - 1 h	1 min	< 0.1 s
Radiation	yes	no	yes	no

Figure 2.1 Comparison of the performance of MPI with established medical imaging modalities [Kno11].

troduce the underlying technology, in a first step, the magnetic properties of the SPIO particles will be discussed in section 2.2 on the tracer's characteristic fingerprint. Subsequently, the methods applied for signal reception (section 2.3), signal encoding (section 2.4) and spatial encoding (section 2.5) will be introduced. Before discussing MPI reconstruction in section 2.7, the role of the MPI system function needs to be described (section 2.6). Since the final goal of this thesis is the implementation of the first dynamic FFL scanner, the signal chain, which is the basis for scanner implementation in MPI, is presented in section 2.8.

2.1 Medical Applications

Regarding medical applications for MPI, several different scenarios arise. Since MPI offers not only one advantage compared to existing medical imaging technologies, the applications vary depending on which of the advantages improves diagnostics or therapy.

The intrinsic property of MPI to provide 3D data might considerably improve cardiovascular interventions. Today, fluoroscopy including intraarterial digital subtraction angiography (DSA) is commonly used in cardiovascular interventional procedures [BBE+12]. The drawbacks of DSA of providing only two dimensional (2D) data and exposing the patient as well as the medical doctor to ionizing radiation is compensated for using MPI, where the vasculature including its pathologies can be visualized in 3D and real-time without the need

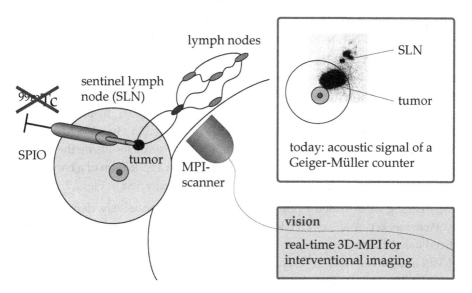

Figure 2.2 A hand-held single-sided MPI scanner will be applied in the SLNB to detect the localization of the sentinel lymph node as well as the lymph nodes connected to it using an SPIO tracer.

of harmful radiation. This way, imaging of stenosis and occlusions using MPI will improve not only diagnostics and therapy planning, but will also offer the possibility of supervision during interventions. Furthermore, the success of the intervention can be verified after the procedure. The navigation of instruments coated with SPIO nanoparticles might also play an important role regarding interventional MPI [HRG+12].

Another very promising scenario is the use of a single-sided MPI scanner [SKB+09] in the SLNB [RBK+09, FRB+10, SEB+12]. The sentinal lymph node (SLN), which is located closest to the tumor in the case of lymphogenous metastasis, is most likely infiltrated by cancer cells. Today, in the SLNB either dye or a radionuclide tracer is used to detect the SLN as well as lymph nodes connected to it, which are also likely to be tumorous and hence need to be removed. When using dye, the physician needs to perform a surgery to trace the dye within the patient, which is with no doubt an invasive procedure. Using a radionuclide like technetium, which is traced by a Geiger-Müller counter, exposes the patient and the physician to harmful radiation.

MPI might provide a solution to perform an SLNB without the need of surgery or radiation for identifying the critical lymph nodes. In this scenario, a hand-held single-sided MPI scanner is used to trace the SPIO nanoparticles, which are injected into the tumor and will from there proceed to the connected lymph nodes. A single-sided MPI scanner provides all components on only one side of the patient and therefore ensures perfect patient access. In the presented scenario, all single-sided MPI components are integrated in a hand-held scanner comparable to an ultrasound probe. The SPIO tracer particles inside the critical lymph nodes will be located using 3D MPI data. The application of a hand-held single-sided MPI scanner supporting the SLNB is illustrated in Fig. 2.2.

The imaging of fast blood flow phenomena was impressively demonstrated by Weizenecker et al. in [WGR+09], where a beating mouse heart was visualized using the first experimental MPI scanner setup. Regarding the progress of MPI in all areas, not only scanner generation, but also concerning reconstruction methods and image processing, the potential held by MPI has only been touched in those experiments. Applications involving fast dynamic blood flow phenomena, as also the catheter laboratory does, will considerably benefit from this new imaging technology especially once MPI has grown to its full potential.

2.2 Fingerprint of the MPI Tracer

MPI is a revolutionary tracer-based medical imaging modality. SPIO nanoparticles are injected into the patients blood circulation, where they can be traced due to their specific magnetization behavior, when exposed to a combination of dynamic and static magnetic fields. When excited, the SPIO nanoparticles exhibit a non-linear change in magnetization, which induces a characteristic voltage signal in receive coils such as a fingerprint.

To understand the origin of the magnetic properties of the nanoparticles enabling their detection and localization, a closer look at their composition as well as at superparamagnetic material in general needs to be taken.

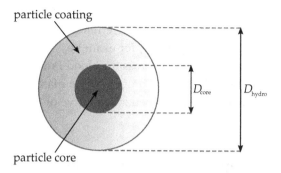

particle coating

D_{core} D_{hydro}

particle core

Figure 2.3 Composition of the SPIO tracer particles. The magnetic core is coated with a biocompatible shell.

2.2.1 Composition of SPIO Nanoparticles

The significance of magnetic nanoparticles regarding medical applications considerably increased within the last decade. Such particles are not only used as tracer in MPI or as contrast agent in MRI, where the magnetic properties of these particles are exploited to image their distribution or to increase image contrast, but also as therapeutic appliance for example in hyperthermia, where magnetic nanoparticles are used for cancer therapy.

Since the magnetic nanoparticles are injected into the patient body, biocompatibility will always be an issue. Fortunately, the clinically approved MRI contrast agent Resovist® [LBF+97] by Bayer Schering Pharma shows significant performance as MPI tracer material. This incident is an important aspect towards clinical acceptability of MPI, since the first *in-vivo* mice images were performed at a clinically approved concentration of Resovist® [WBG07].

Regarding the composition of magnetic nanoparticles, two aspects are of importance: the magnetic core and the non-magnetic coating. Relevant for the imaging process are the magnetic properties of the core material of the particle. However, the characteristic magnetization behavior only arises for non-interacting particles. Without an appropriate coating, the particles would cluster due to their magnetic properties and would not exhibit superparamagnetic behavior. The coating also ensures biocompatibility of the particles. The schematic structure of an MPI tracer particle is illustrated in Fig. 2.3. The hydrodynamic diameter D_{hydro} of the particle including its coating is an important property with respect to medical applications, because it determines the overall size of the particle. The overall size gains importance, for example, when considering applications, in which the particles need to cross the blood-brain barrier. The important property concerning the imaging process is the diameter of the magnetic core D_{core}, since the shape of the magnetization curve of the particles strongly depends on this parameter.

2.2.2 Phenomenology of Superparamagnetism

A simple but very common model describing the magnetization behavior of SPIO nanoparticles is Langevin's theory of paramagnetism. However, the magnetic core of the particles consists of a ferromagnetic substance. Why use a theory originally developed to describe paramagnetic material? To answer this question, it is necessary to examine the magnetic properties of ferromagnetic materials.

Ferromagnetic materials are characterized by the formation of magnetic domains, named Weiss domains, in which the atomic magnetic moments μ_a are aligned resulting in a non-zero magnetization. Even in the absence of an external magnetic field, Weiss domains are present. Exposing a ferromagnetic material to an external magnetic field will parallelize the direction of the orientation of the atomic magnetic moments μ_a of all domains. The size of these domains depends on the corresponding material. For spherical particles, the critical diameter D_{cr}, which is identical to the largest diameter being energetically efficient for a particle to consist of one single domain [Kro07], can be calculated by

$$D_{cr} = \frac{72 \sqrt{AK_A}}{\mu_0 M_{sat}^2},$$ (2.1)

where K_A is the anisotropy constant and A is the exchange stiffness constant. The critical diameter D_{cr} only depends on material constants [Rog11]. In MPI, magnetite Fe_3O_4 is most often used as tracer and the corresponding saturation field strength of magnetite is given by [KB12]

$$M_{sat} = 0.6 \, T \mu_0^{-1}.$$ (2.2)

The critical diameter of magnetite is not consistently specified in literature. Values from 12.4 nm [Kro07] up to a range of 30 - 100 nm [RMC03] are mentioned

[Rog11]. The MPI tracer particles used today range from some nanometers up to not more than 30 nm, where the larger particles are present in a much lower density than the smaller ones. Due to this reason, we will for now assume a one domain structure of the SPIO tracer particles. Hence, all aligned atomic magnetic moments μ_a of each particle add up to one (*super*) large particle magnetic moment μ_p.

Considering a distribution of such one domain nanoparticles, another property is of importance: the non-magnetic coating. As discussed in the previous section, the coating ensures biocompatibility of the tracer. In addition, however, it also prevents agglomeration of the particles. Adequately coated, the SPIO particles may therefore in good approximation be assumed as non-interacting. The behavior of the magnetic moments μ_p of the SPIO particles can hence be compared to the behavior of the non-interacting atomic magnetic moments μ_a of a paramagnetic material. This instance is the historical origin of the name *superparamagnetism*, which describes the paramagnetic behavior of the much larger *super* particle magnetic moments μ_p of the SPIO nanoparticles. And exactly for this reason, a theory describing paramagnetism can be applied for nanoparticles made of a ferromagnetic substance [BL59].

2.2.3 Magnetic Moment

Regarding a distribution of SPIO particles in a stable suspension, i.e. a ferrofluid, the magnetic moment of a single particle depends on the volume of the particle core V_c as well as on the bulk saturation magnetization of the core material M_{sat} (Eq. (2.2)) [Rog11]

$$\mu = V_c M_{\text{sat}} \hat{e}_\mu, \tag{2.3}$$

where $\hat{e}_\mu$ denotes a unit vector in direction of the orientation of the magnetic moment of the SPIO particle.

However, this equation may - as the assumption of considering purely one domain particles - only be regarded as an approximation, since it describes the case of perfectly aligned atomic magnetic moments μ_a in the one magnetic domain of the corresponding nanoparticle. The magnetic moment of an SPIO particle, however, depends on the exact formation of the magnetic domains, which will in reality not always be ideally homogeneous. For different particle sizes, inhomogeneities in the magnetic alignment may occur. For particle sizes below 3 nm, the intrinsic force of the particle will not suffice to cause ferromagnetic behavior [GSV06]. Furthermore, surface effects and the formation of vortices occurring for particles with a diameter close to D_{cr} will also influence domain formation [Gui09, Rog11].

2.2.4 Magnetization

The significant property of SPIO particles regarding image generation in MPI is their characteristic non-linear magnetization behavior. Magnetization is generally defined as the magnetic moment per volume

$$\mathbf{M} := \frac{\mathrm{d}\mu}{\mathrm{d}V} \tag{2.4}$$

and is a physical quantity defined for a distribution of magnetic moments. Considering a small finite volume ΔV, which is of the size of an image voxel, and N being the number of SPIO particles in this volume, a discrete version of Eq. (2.4) can be derived

$$M = \frac{1}{\Delta V} \sum_{j=1}^{N} \mu_j, \tag{2.5}$$

where μ_j denotes the magnetic moment of the jth SPIO particle. From now on we will describe the magnetic behavior of SPIO particles including all assumptions introduced in the previous sections. The subscript p has therefore been omitted and μ will denote the particles magnetic moment in the following.

For a distribution of SPIO particles, it is convenient to introduce a mean magnetic moment

$$\bar{\mu} = \frac{1}{N} \sum_{j=1}^{N} \mu_j. \tag{2.6}$$

When defining the SPIO particle concentration c as the number of particles N per volume ΔV, the magnetization of the particles in Eq. (2.5) can be reformulated using Eq. (2.6) as

$$M = \frac{N}{\Delta V}\bar{\mu} = c \cdot \bar{\mu}. \tag{2.7}$$

Hence, a linear connection between the particle magnetization M and the concentration of SPIO particles c in a considered volume unveils. As will be discussed in section 2.3, the signal measured in MPI receive coils is proportional to the derivative of the particle magnetization with respect to time. Since the derivative is a linear operation, the connection of the MPI receive signal and the particle concentration can also be described in a linear manner.

When applying an external magnetic field H to a distribution of SPIO particles, the magnetic moments of the particles tend to align within the direction of H resulting in a non-zero magnetization. At a specific field strength H_{sat}, all magnetic moments are aligned within the direction of H so that further increasing the field strength will not increase the magnetization. This mechanism is illustrated in Fig. 2.4 and results in a non-linear magnetization behavior. At zero field strength, the magnetic moments of the particles are thermally distributed and the magnetization is equal to zero (petrol region). When increasing the field strength, the particles begin to align within the direction of the field and cause a non-zero magnetization (turquois region). Increasing the field strength beyond H_{sat}, however, will not lead to a further increase in magnetization. The particle distribution is magnetically saturated (gray region).

Precisely this non-linear magnetization behavior is the fundamental characteristic of a superparamagnetic material enabling their detection using MPI. Neglecting anisotropy of the particles as well as hysteresis effects, the magnetization behavior of a distribution of SPIO particles can be described using Langevin's theory of paramagnetism, which will be derived in the subsequent section.

2.2.5 Langevin Theory of Paramagnetism

Langevin's theory of paramagnetism describes the non-linear magnetization behavior of SPIO tracer particles if anisotropy and hysteresis effects are neglected. The presented derivation is based on [CC64].

If an external magnetic field H is applied to a distribution of non-interacting magnetic moments μ, a couple of force $-\mu H \sin\theta$ will act on each of the particle's magnetic moments. The angle between the external magnetic field H and the direction of the magnetic moment μ is denoted θ. This force is responsible for the tendency of the particles to align in direction of the external magnetic

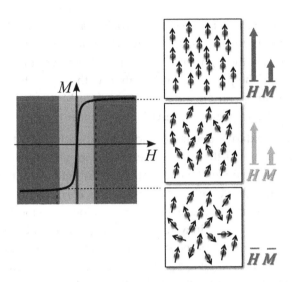

Figure 2.4 The effect of saturation of a distribution of SPIO nanoparticles. At no external magnetic field H, the magnetic moments of the SPIO particles are thermally distributed and no magnetization arises. When applying an external magnetic field H, the particles' magnetic moments start to align within the direction of H. At the saturation field strength, all magnetic moments of the particles are aligned along the direction of the external magnetic field H and an increase in field strength will not result in an additional increase in magnetization. Figure based on [KB12].

field, which does finally lead to the phenomenon of magnetic saturation once all magnetic moments are parallelized.

For a finite temperature, thermal agitation counteracts this tendency and leads to a potential energy

$$E_p = -\mu_0 \mu H \cos \theta \tag{2.8}$$

of each moment μ, when defined according to Eq. (2.3) with the bulk saturation magnetization given by Eq. (2.2).

The total number of magnetic moments in a unit volume, which are rotated by an angle between θ and $\theta + d\theta$ with respect to the orientation of the external magnetic field H needs to be proportional to the solid angle $2\pi \sin \theta\, d\theta$ as well as to the Boltzmann factor [CC64].

Considering a particle at temperature T_P in thermal equilibrium, the Boltzmann factor describes the relative probability $P(E)$ to find a particle in state E and is given by

$$P(E) \propto \exp\left\{-\frac{E}{k_B T_P}\right\}. \tag{2.9}$$

Since the rotation of the magnetic moment μ with respect to the orientation of H leads to a potential energy $E_p = -\mu_0 \mu H \cos\theta$ of μ, the Boltzmann factor is in the considered case given by

$$P(E) = \exp\left\{\frac{\mu_0 \mu H \cos\theta}{k_B T_P}\right\}. \tag{2.10}$$

Hence, the total number of magnetic moments in a unit volume, which are rotated by an angle between θ and $\theta + d\theta$ with respect to the orientation of the external magnetic field H is given by

$$n(\theta)\,d\theta = 2\pi n_0 \exp\left\{\frac{\mu_0 \mu H \cos\theta}{k_B T_P}\right\} \sin\theta\,d\theta. \tag{2.11}$$

Again assuming N to be the total number of magnetic moments per unit volume,

$$\int_0^\pi n(\theta)\,d\theta = 2\pi n_0 \int_0^\pi \exp\left\{\frac{\mu_0 \mu H \cos\theta}{k_B T_P}\right\} \sin\theta\,d\theta = N \tag{2.12}$$

is fulfilled.

The magnetization of a system of SPIO particle magnetic moments in a volume V is determined using [CC64]

$$M(H,t) = \frac{1}{V} \int_0^\pi \mu \cos\theta\, n(\theta)\,d\theta. \tag{2.13}$$

Inserting the findings expressed in Eq. (2.12) and expanding with N reveals

$$M(H,t) = \frac{N\mu}{V} \frac{\displaystyle\int_0^\pi n(\theta)\cos\theta\,d\theta}{\displaystyle\int_0^\pi n(\theta)\,d\theta} \tag{2.14}$$

$$= \frac{N\mu}{V} \frac{\int_0^\pi \exp\left\{\frac{\mu_0\mu H \cos\theta}{k_B T_P}\right\} \sin\theta\cos\theta\, d\theta}{\int_0^\pi \exp\left\{\frac{\mu_0\mu H \cos\theta}{k_B T_P}\right\} \sin\theta\, d\theta}.$$

A substitution of $\mu_0\mu H/k_B T_P = \alpha$ and $\cos\theta = x$ is performed leading to

$$M(H,t) = \frac{N\mu}{V} \frac{\int_1^{-1} e^{\alpha x} x\, dx}{\int_1^{-1} e^{\alpha x}\, dx} \tag{2.15}$$

$$= \frac{N\mu}{V} \frac{\frac{1}{\alpha}(e^\alpha + e^{-\alpha}) - \frac{1}{\alpha^2}(e^\alpha - e^{-\alpha})}{\frac{1}{\alpha}(e^\alpha - e^{-\alpha})}$$

$$= \frac{N\mu}{V}\left(\frac{e^\alpha + e^{-\alpha}}{e^\alpha - e^{-\alpha}} - \frac{1}{\alpha}\right)$$

$$= \frac{N\mu}{V}\left(\coth\alpha - \frac{1}{\alpha}\right).$$

The magnetization behavior of an SPIO particle distribution can hence be described by means of the Langevin function

$$\mathcal{L} = \left(\coth\alpha - \frac{1}{\alpha}\right), \tag{2.16}$$

and is given by

$$M(H,t) = \frac{N\mu}{V}\mathcal{L}(\alpha) \tag{2.17}$$

with $\alpha = \mu_0\mu H/k_B T_P$.

The shape of the magnetization curve of a distribution of SPIO particles depends on the core diameter of the particles via the magnetic moment μ. The magnetization behavior of a distribution of monodisperse SPIO particles according to Langevin's function is plotted for different particle diameters in Fig. 2.5. For ideal particles, the magnetization behaves according to a step function.

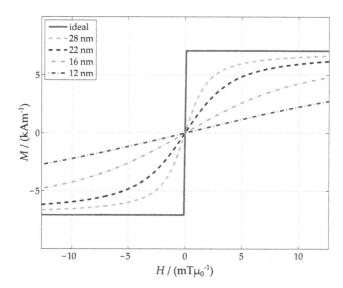

Figure 2.5 Magnetization behavior of a distribution of SPIO particles according to the Langevin theory for various particle diameters.

2.2.6 Distribution of SPIO Nanoparticle Core Diameters

During a synthesis process, SPIO particles with a variance of physical properties are produced. Ideally, monodisperse, isotropic particles optimized with respect to their performance in MPI should be synthesized. Unfortunately, this could not yet be realized. The outcome of synthesis processes rather is a distribution of particles of various core diameters. In a conventional synthesis process the diameters of the SPIO cores is log-normal distributed [KSNG99]. The probability of a specific core diameter to appear is given by

$$
\rho\left(D_{\text{core}}\right) = \begin{cases} \dfrac{1}{\sigma D_{\text{core}}\sqrt{2\pi}} \exp\left\{-\dfrac{1}{2}\left(\dfrac{\ln\left(D_{\text{core}}\right) - \mu}{\sigma}\right)^{2}\right\} & , \text{if } D_{\text{core}} > 0 \\ 0 & , \text{if } D_{\text{core}} \leq 0 \end{cases}, \quad (2.18)
$$

where μ and σ are parameters of the log-normal distributed core diameter. They can be expressed in terms of the expectation value $E(D_{\text{core}})$ and the variance $\text{Var}(D_{\text{core}})$ by [BKS+09, Bie12]

$$
\mu = \ln\left(E(D_{\text{core}})\right) - \frac{1}{2}\ln\left(\frac{\text{Var}(D_{\text{core}})}{E^{2}(D_{\text{core}})} + 1\right) \quad (2.19)
$$

and

$$\sigma = \sqrt{\ln\left(\frac{\mathrm{Var}(D_{\mathrm{core}})}{\mathrm{E}^2(D_{\mathrm{core}})} + 1\right)}. \qquad (2.20)$$

2.2.7 Anisotropy Effects

The presented descriptions of the magnetic properties of SPIO particles include several assumptions. First of all, only one-domain particles with perfectly homogeneous aligned atomic magnetic moments are considered. Hence, magnetic anisotropy has been omitted. Secondly, only spherical particles are included into the model and shape anisotropy is neglected.

In a realistic ferrofluid, however, both anisotropies will be present and will lead to a delayed alignment of the particle's magnetic moment when exposed to an external magnetic field. This results in a finite relaxation time of the particles. Two different relaxation processes are present regarding an SPIO nanoparticle suspension. The first process, named Brownian rotation, describes a physical rotation of the particle including its magnetic moment (upper row of Fig. 2.6). The second relaxation phenomenon is caused by the rotation of the magnetic moment of a particle with respect to its physical axis and is called Néel rotation (lower row of Fig. 2.6).

When relaxation occurs, the magnetic moment of an SPIO particle does not instantaneously follow the direction of the external magnetic field. In turn, applying a static magnetic field to a distribution of SPIO particles will lead to a remanent magnetization even after deactivating the field. The magnetization of the system will decrease not instantaneously, but exponentially according to

$$M(t) = M_0 \exp\left\{\frac{t}{\tau}\right\}. \qquad (2.21)$$

Here, the absolute value of the remanent magnetization with respect to time is denoted $M(t)$, M_0 is the magnetization of the system at the time of field deactivation and τ is the relaxation time [KB12].

Regarding MPI, relaxation effects need to be taken into account, if the frequency f_D of the drive field is in the rage of the inverse of the relaxation time τ. Ideally, the drive field frequency should be much lower than the inverse relaxation time

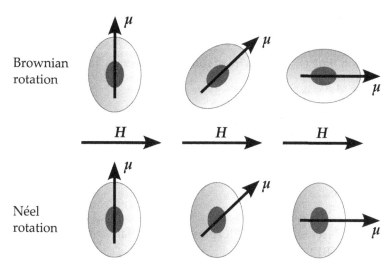

Figure 2.6 The Néel and the Brownian relaxation of SPIO particles. The Brownian relaxation describes the effect arising from the physical rotation of the particle with a fixed direction of its magnetic moment (upper row). Néel relaxation in contrast describes the rotation of the magnetic moment itself, while the physical axis of the particle stays fixed. Figure based on [KB12].

$$f_D \ll \frac{1}{\tau}. \tag{2.22}$$

In that case, the magnetic moments of the particles will be able to follow the direction of the dynamic MPI drive field.

Hence, the relaxation time is the central quantity determining whether the Langevin model introduced in section 2.2.5 sufficiently describes the SPIO particle behavior. The relaxation times connected Brownian and Néel rotation processes shall be introduced in the following.

BROWNIAN ROTATION

The relaxation time for the Brownian rotation is given by [Bro63, KB12]

$$\tau_{\text{Brown}} = \frac{3\eta V_{\text{hydro}}}{k_B T_P}. \tag{2.23}$$

It linearly scales with the hydrodynamic volume V_{hydro} of the SPIO particle and the viscosity of the ferrofluid, denoted η.

NÉEL ROTATION

Regarding the Néel rotation of SPIO particles, the relaxation time is determined by [N49, N55, KB12]

$$\tau_{\text{Néel}} = \tau_0 \exp\left\{ \frac{K_A V_{\text{core}}}{6 k_B T_P} \right\}. \tag{2.24}$$

It exponentially depends on the anisotropy constant of the ferrofluid K_A as well as on the volume of the magnetic core V_{core} of the SPIO particle. The constant factor τ_0 is in the order of $(10^{-11} - 10^{-9})\,\text{s}^{-1}$, which corresponds to the frequency of gyromagnetic rotation [Rog11].

The exponential increase of the Néel relaxation time for large magnetic core volumes can be explained by a deformation of the magnetic domains. The spins will no longer be parallel such that the assumption of considering purely one-domain particles will no longer be valid. Due to these effects, the particles are said to be blocked for exceeding a certain blocking volume

$$V_{\text{core}}^{\text{block}} = \ln\left(\frac{\tau_m}{\tau_0} \right) \frac{k_B T_P}{K_A}, \quad \text{for } T_P = \text{constant}, \tag{2.25}$$

or blocking temperature, respectively

$$T_P^{\text{block}} = \ln\left(\frac{\tau_m}{\tau_0} \right) \frac{K_A V_{\text{core}}}{k_B}, \quad \text{for } V_{\text{core}} = \text{constant}. \tag{2.26}$$

In these relations, τ_m denotes a characteristic measurement time [Gui09, Rog11].

COMBINED ROTATION

The combined rotation including Brownian and Néel rotation processes is described by an efficient relaxation time [SS94]

$$\tau_{\text{eff}} = \frac{\tau_{\text{Brown}}\tau_{\text{Néel}}}{\tau_{\text{Brown}} + \tau_{\text{Néel}}}. \tag{2.27}$$

Hence, the process with the shortest of the two relaxation times will dominate the combined rotation. Regarding the particle volume, three groups can be identified

1. small particles will most likely exhibit Néel rotation,

2. particles of intermediate volume will perform a combined rotation, and

3. large particles will be dominated by the Brownian rotation due to the blocked Néel rotation [Rog11].

To include anisotropy effects, a more advanced particle model is needed to fully understand the behavior of SPIO tracer particles in MPI. However, it has been demonstrated in [KBS+09] and [KBS+10b] that the simple particle model using Langevin's function leads to acceptable results when included it into model-based reconstruction algorithms. Therefore, anisotropy effects will be neglected in this thesis. For a more detailed particle model including a description of anisoptropy effects of SPIO particles, additional resource is recommended [Rog11].

2.3 Signal Reception

MPI enables imaging of the inside of the human body due to the interaction of the specific magnetization behavior of the SPIO tracer particles, which has been discussed in the previous section, and a combination of static and dynamic magnetic fields. These fields will be introduced in the subsequent sections 2.4 and 2.5. In a first step, however, a method for detecting the characteristic fingerprint of the particles needs to be derived. Therefore, MPI signal reception shall be introduced in the following.

A change in magnetization, as present when exposing a distribution of SPIO particles to an external magnetic field H, induces a voltage u in receive coils according to Faraday's law [Max73]

$$u(t) = -\mu_0 \int_{\Omega} \frac{d}{dt} M\left(H(r,t), r, t\right) p(r) \, d^3r. \tag{2.28}$$

In this formula, the coil sensitivity of the receive coil $p(r)$ is introduced, which equals the proportionality factor between the magnetic field strength H and the coil current I and is hence given by

$$p(r) = \frac{1}{4\pi} \int_{\mathbb{R}^3} \frac{j(r') \times (r - r')}{|r - r'|^3} d^3r' = \frac{H(r)}{I}. \tag{2.29}$$

Coil sensitivities will be of importance in chapter 6, where an optimal choice of currents is sought for a given magnetic target field H_{target}. For a specific coil or more general conducting medium, the sensitivity can be predetermined when simulating the magnetic field $\hat{H}$ generated by the coil or conducting medium at a given current $\hat{I}$ using

$$p(r) = \frac{\hat{H}(r)}{\hat{I}}. \tag{2.30}$$

Solving the minimization problem

$$\arg\min_{I} \|Ip - H_{target}\|_2^2 \tag{2.31}$$

will then result in an optimal current I to generate the desired target field H_{target}.

Taking a closer look at the voltage induced in the receive coil due to the SPIO particles' change in magnetization as a reaction on an external magnetic field given in Eq. (2.28), it is possible to identify characteristic conditions on the magnetization curve of the SPIO tracer material as well as on the external magnetic field, which will positively influence the MPI signal.

In a first step, equation (2.28) will be simplified assuming the following conditions

1. the external magnetic field H and the receive coil sensitivity p are homogeneous over the volume of interest V,

2. the external magnetic field H and the receive coil sensitivity p are parallel over V, hence, the magnetization M will also be parallel to the receive coil sensitivity p,

3. M denotes the magnetization component picked up by the receive coil, and

4. the particle distribution is approximated by a δ-distribution.

With these assumptions, equation (2.28) can be simplified to [Rog11, RWGB09]

$$u(t) = -\mu_0 p \frac{\mathrm{d}}{\mathrm{d}t} M\left(H(t)\right). \tag{2.32}$$

Neglecting constant factors for simplicity leads to

$$u(t) = -\frac{\mathrm{d}}{\mathrm{d}t} M\left(H(t)\right) \tag{2.33}$$

$$= -\frac{\partial M}{\partial H} \frac{\mathrm{d}H(t)}{\mathrm{d}t}. \tag{2.34}$$

From this formula, two important aspects influencing the magnitude of the signal detected in MPI are visible. To achieve a high signal

1. the magnetization M needs to change fast with respect to the external magnetic field H and

2. the external magnetic field H needs to change fast with respect to time t.

Hence, a steep magnetization curve as well as a high excitation frequency are essential for signal quality in MPI [RWGB09].

2.4 Signal Encoding - The Drive Field

The characteristic magnetization behavior of the SPIO particles constitutes the substantial physical phenomenon enabling their detection using MPI. Since this magnetization behavior occurs as a reaction on external magnetic fields, these fields need to be discussed in detail.

In MPI two different magnetic fields are applied to realize the imaging process, the *drive field* and the *selection field*. The spatially homogeneous, time varying drive field is responsible for the excitation of the SPIO particles, which results in a magnetization change, which induces a characteristic voltage signal in receive coils. This characteristic signal gives evidence for the existence of SPIO particles within the considered volume. Hence, the drive field enables detection of SPIO tracer particles, i.e. signal encoding.

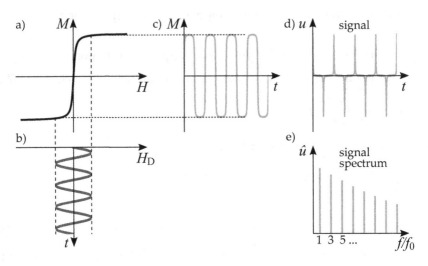

Figure 2.7 MPI signal encoding. A distribution of SPIO particles with non-linear magnetization behavior according to the Langevin function a) is excited by a sinusoidal drive field b). The resulting magnetization over time c) corresponds to a modulated sine wave. The signal spectrum e) of the derivative of the magnetization, which is measured in receive coils d), contains higher harmonics of the drive field frequency f_D. Figure based on [RWGB09, Kno11].

In this section the drive field will be introduced in detail, while the selection field will be discussed in the subsequent section 2.5.

The MPI drive field conventionally is of sinusoidal shape

$$H_D = A_D \sin(2\pi f_D t). \tag{2.35}$$

The drive field amplitude is denoted A_D, while f_D is the drive field frequency. The sinusoidal time varying drive field enforces a change onto the magnetization of an SPIO particle distribution. This is illustrated in Fig. 2.7, where the black curve in Fig. 2.7 a) shows the magnetization behavior of a distribution of SPIO particles with respect to an external magnetic field as described by the Langevin function, Fig. 2.7 b) shows the drive field amplitude with respect to time, and the resulting developing of the SPIO particle magnetization with time is illustrated in Fig. 2.7 c). Due to the non-linear magnetization behavior of the SPIO particles, which occurs if A_D is high enough to cover the dynamic region of the particles' magnetization curve, the change in magnetization corresponds to a modulated sine and the Fourier series of the measured voltage $u(t)$ contains higher harmonics of the excitation frequency f_D

$$u(t) = \sum_{n=-\infty}^{+\infty} \hat{u}_n e^{in\omega t}. \tag{2.36}$$

With the period $T = 1/f_D = 2\pi/\omega$, the Fourier coefficients are calculated via

$$\hat{u}_n = \frac{1}{T} \int_0^T u(t) e^{-in\omega t} \, dt. \tag{2.37}$$

For a material with a linear magnetization behavior, the magnetization answer on a sinusoidal excitation field would also be of sinusoidal shape. Hence, the Fourier spectrum of the measured voltage would only contain the excitation frequency f_D. Since the SPIO particles, however, exhibit the characteristic non-linear magnetization behavior, existence of higher harmonics in the Fourier spectrum of the measured voltage signal gives evidence for the existence of SPIO particles inside of the measured volume.

Regarding scanner implementation, the MPI drive field needs to be generated by a specific assembly of electromagnetic coils. Such a spatially homogeneous magnetic field can be generated by a Helmholtz coil pair. This is a coil pair with equal properties of both coils, equal currents orientated parallel to each other an in an ideal case a characteristic distance of the coils equal to their radius. An exemplary Helmholtz coil pair generating an ideal homogeneous magnetic field, as desirable for the MPI drive field, is illustrated in figure 2.8.

When measuring the derivative of the magnetization response in a volume of interest using receive coils, the existence of SPIO tracer material within this volume correlates with the appearance of higher harmonics of the drive field frequency f_D in the signal spectrum $\hat{u}$. At this point, SPIO tracer particles can be detected using MPI. However, all particles within the considered volume contribute to the measured signal and can not yet be spatially resolved. To achieve spatial encoding, the superposition of the MPI selection field is needed, which will be introduced in the subsequent section.

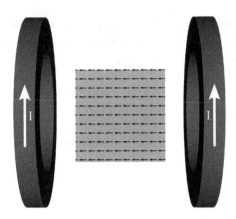

Figure 2.8 Helmholtz coil pair generating a spatially homogeneous magnetic field in a region around the center of the coil pair.

2.5 Spatial Encoding - The Selection Field

Up to this point, all particles in the considered volume are detected simultaneously. It is hence not possible to distinguish between particles at different spatial positions. To achieve spatial encoding, the MPI *selection field* is introduced, which enables spatial selection of a certain region, in which particles contribute to the measured signal.

In contrast to the drive field, the selection field is a static magnetic field featuring an FFP or an FFL in the center and a linearly increasing magnetic field strength in all spatial directions. Such a magnetic field is called constant gradient field. The selection field *selects* a specific region in which particles contribute to the measured signal and is therefore responsible for spatial encoding.

To understand the process of spatial encoding, the magnetization behavior of the SPIO particles exposed not only to the drive field, but also to a superimposed constant offset field needs to be discussed. If the offset field strength is chosen large enough to drive and keep the SPIO particles in a state of magnetic saturation, even if an additional drive field is superimposed, no change in the particles' magnetization will occur. Hence, even in the presence of SPIO particles within the measured volume and despite applying the MPI drive field to the SPIO particle distribution, no particle signal will be detected by the receive

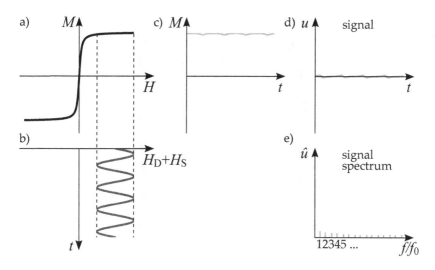

Figure 2.9 MPI signal generation for a constant offset field superimposed to the drive field. The working point on the magnetization curve is shifted depending on the offset field strength. In the presented case, the offset field strength is high enough to keep the SPIO particles in a state of saturation. The change of the magnetization over time is hence minimal c) and in turn, almost no signal will be detected in the receive coil d) and no signal spectrum is obtained e). Figure based on [RWGB09, Kno11].

coils. This situation is illustrated in Fig. 2.9, where the signal generation process for the MPI drive field with a superimposed constant offset field is plotted.

To extend these considerations from a constant offset field to the constant gradient MPI selection field, a closer look at the selection field shape needs to be taken. Due to Maxwell's equations, or Gauss's law for magnetism in particular

$$\nabla \cdot H = 0, \tag{2.38}$$

the gradient strength of the magnetic field will never be equal in all three spatial directions. It will be elliptically shaped with a gradient strength of g in one direction and $-\frac{1}{2}g$ in the remaining two directions

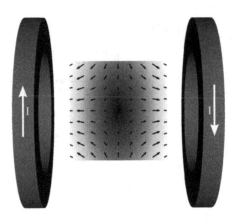

Figure 2.10 FFP field generated by a Maxwell coil pair.

$$H_S = g \begin{pmatrix} 1 & 0 & 0 \\ 0 & -\frac{1}{2} & 0 \\ 0 & 0 & -\frac{1}{2} \end{pmatrix}. \tag{2.39}$$

A way of realizing FFP field generation is the use of a Maxwell coil pair. Similar to the Helmholtz coil pair, the properties of the two coils are identical and their distance ideally resembles their radius, only that the current is orientated antiparallel. Fig. 2.10 illustrates the generation of an FFP field using such a Maxwell coil pair.

Only particles at or in the close vicinity of the FFP are free to change the orientation of their magnetic moments. Hence, only particles at these points will undergo a magnetization change, when excited by the drive field. Particles of higher distance to the FFP are driven into saturation due to the increasing field strength of the selection field. Exactly this method is applied to achieve spatial encoding. The FFP is steered through the field of view and only particles in the vicinity of the FFP as well as at the FFP will contribute to the measured signal.

In detail, not all particles at and around the FFP exhibit the same magnetic behavior. Their magnetization characteristics depend on the strength and shape of the selection field. Each particle distribution at a specific area will lead to a specific signal. This fact will be included in the reconstruction method and is illustrated in Fig. 2.11 for three different spatial positions and in turn selection field characteristics.

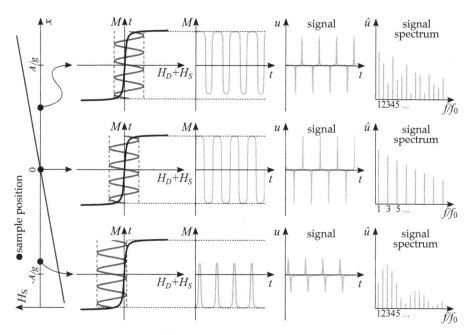

Figure 2.11 Spatial encoding in MPI regarding different spatial positions. The signal generated by a distribution of SPIO particles strongly depends on the external magnetic fields applied to it. Changing the working point on the magnetization curve of the particle distribution will also change the induced voltage and the signal spectrum of the same. To differentiate the signal distributions of particles located at different spatial positions with respect to the FFP, a reconstruction step is needed, which will be introduced in section 2.7. Figure based on [RWGB09, Kno11].

The shape of the selection field will play an important role in the course of this thesis. Magnetic field generation regarding scanner implementation needs to be realized either by electromagnetic coils or permanent magnets. Since the selection field is of static nature in the case of FFP imaging, it is possible to use permanent magnets for selection field generation. This was also done in the first MPI scanner presented in [GW05]. In this thesis, however, an alternative selection field shape, i.e. a field featuring an FFL instead of an FFP, is presented. The FFL selection field is characterized by a line of zero field strength in its center and a constant gradient perpendicular to this line. Due to the line characteristic,

Gauss's law for magnetism allows for equal but contrarily orientated field gradients in the remaining two directions. The FFL selection field is hence given by

$$\boldsymbol{H}_S = g \begin{pmatrix} 0 & 0 & 0 \\ 0 & -1 & 0 \\ 0 & 0 & 1 \end{pmatrix} \qquad (2.40)$$

and is illustrated in Fig. 2.12. Magnetic field generation for FFL imaging will be discussed in detail in section 4.

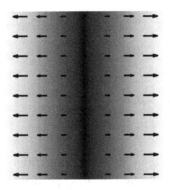

Figure 2.12 An FFL selection field.

One important difference of FFL imaging compared to FFP imaging shall be mentioned at this point: the selection field is no longer static, but rotates with a certain frequency f_S. For realizing this rotation, no mechanical movement shall be used to guarantee real-time capability of the FFL imaging device. The rotation will be realized by varying the currents in electromagnetic field generating coils. Hence, it is not possible to use only permanent magnets for FFL selection field generation. It will be shown in section 4.3 that the selection field can be separated into a static and a dynamic part [KEB+10]. The static part can be generated by permanent magnets to lower electrical power consumption. The dynamic part, however, needs to be generated by electromagnetic coils.

The shape of both, permanent magnets as well as electromagnetic coils strongly influences the quality of the generated magnetic field. Considering a fixed field of view (FOV) size, the coil pair generating either a homogeneous field or a gradient field should be chosen as large and far away and as thin as possible,

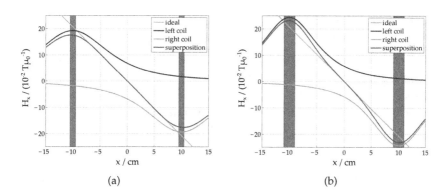

(a)   (b)

Figure 2.13 Magnetic field generated by an ideally thin (a) and a realistic (b) Maxwell coil pair. The deviation from an ideal gradient field rises in case of considering realistic coil properties as in case (b).

to ensure homogeneity or gradient linearity of the field, respectively. In reality, however, this is not feasibly due to the increase in current density for thinner coils and the increase in electrical power consumption for coils further away from the FOV needed to generate a given target field. A compromise between field homogeneity or gradient linearity and current density and electrical power consumption will always be sought.

To illustrate the influence of the coil shape on the gradient linearity of an FFP field, two different Maxwell coil pairs generating a gradient field strength of $2 \, \text{Tm}^{-1}\mu_0^{-1}$ are considered. The radius of both coil pairs is chosen to be 10 cm. The first coil pair is chosen to be ideally thin, while the second coil pair is of realistic shape (thickness of 4 cm and length of 2 cm). The magnetic field generated by the ideally thin coil pair on its axis, i.e. the x-axis is plotted in Fig. 2.13(a). In contrast to these results, Fig. 2.13(b) shows the magnetic field on the axis of the second, more realistic coil pair. The deviation from an ideal gradient field is clearly visible. To ensure high selection field quality, the shape of the coils needs to be optimized. Methods on coil shape optimization will be presented in chapter 6.

2.6 The System Function

In this section, the MPI signal equation shall be derived and the MPI system function is introduced. There are two ways of expressing the MPI signal, in time space and in frequency space. Both methods are shortly presented and discussed in the following. For a more detailed description of the properties of the MPI system function see [RWGB09, RWGB12] and [KB12].

2.6.1 System Function in Time Space

As stated in Eq. (2.7), the magnetization of a distribution of non-interacting SPIO particles can be expressed as a linear relation between the particle concentration c and the mean magnetic moment of the particle distribution $\bar{\mu}$

$$M(H(r,t),r,t) = c(r)\bar{\mu}(H(r,t),t). \tag{2.41}$$

Inserting this expression into Faraday's law, given in Eq. (2.28), leads to the continuous MPI signal equation

$$u(t) = \int_\Omega s(r,t)c(r)\,\mathrm{d}^3r, \tag{2.42}$$

where

$$s(r,t) = -\mu_0 p(r)\frac{\partial\mu(H(r,t),t)}{\partial t} \tag{2.43}$$

is called the system function. Hence, the relation between the MPI signal $u(t)$ and the particle concentration c is linear and can be expressed via an integral transform with the system function $s(r,t)$ being the integral kernel.

The signal equation (2.42) can be expressed as convolution, if this equation is shift invariant [KB12]. For a shift invariant system,

$$\int_\Omega s(r,t)c(r)\,\mathrm{d}^3r = \int_\Omega s(r'(t) - r)c(r)\,\mathrm{d}^3r \tag{2.44}$$

would be fulfilled, where $r'(t)$ is a time dependent spatial position, and the signal equation could hence be expressed as

$$u(t) = \int_\Omega s(r'(t) - r)c(r)\,\mathrm{d}^3r = (s * c)(r'(t)). \tag{2.45}$$

There is one major advantage of formulating the imaging equation in terms of a convolution: efficient reconstruction methods via deconvolution arise. It has been demonstrated in [Sch10] and [GC11] that even the 3D MPI signal equation can be expressed as convolution. However, this is only true when assuming ideal magnetic fields, which will not be provided in a realistic imaging system. Furthermore, deconvolution will only be possible when using the complete voltage signal induced by the particles' change in magnetization. As will be described in section 2.8, the complete signal will not be accessible due to the need of filtering. The particle signal at the excitation frequency cannot be provided. A last aspect derogating shift invariance is the behavior of the SPIO tracer particles, which will in reality show finite relaxation times, which are not yet included into the model [KB12]. Due to these aspects, Eq. (2.45) can only be considered as an approximation. A more realistic and precise way of describing the MPI signal process is given by Eq. (2.42).

2.6.2 System Function in Frequency Space

A common way of describing the MPI signal equation is transforming it into frequency space. The major criterion for using a frequency space representation of the MPI signal equation is the non-accessibility of the particle signal at the excitation frequency [KB12]. As introduced in equation (2.36), the induced voltage can be expressed as Fourier series with coefficients given in equation (2.37). Proceeding as for the derivation of the signal equation in time space, equation (2.41) describing the particle magnetization and Faraday's law equation (2.28) are inserted into equation (2.37) leading to

$$
\hat{u}_n = \int_\Omega c(r) \left(-\frac{\mu_0 \omega}{2\pi} \right) \int_0^T p(r) \frac{\partial}{\partial t} \bar{\mu}\left(H(r,t),t\right) e^{-in\omega t}\, \mathrm{d}t\, \mathrm{d}^3 r. \tag{2.46}
$$

The linearity of the signal equation is not disturbed due to the linearity of the Fourier transformation and the signal equation in frequency space is hence given by

$$
\hat{u}_n = \int_\Omega \hat{s}_n(r) c(r)\, \mathrm{d}^3 r, \tag{2.47}
$$

where the system function in frequency space reads

$$\hat{s}_n(r) = -\frac{\mu_0 \omega}{2\pi} \int_0^T p(r) \frac{\partial}{\partial t} \bar{\mu}\left(H(r,t),t\right) e^{-in\omega t} \, \mathrm{d}t. \tag{2.48}$$

2.6.3 Measurement-based System Function

Today, no exact physical model exists, which sufficiently describes the properties of the MPI scanner as well as the behavior of the SPIO tracer particles. Due to that, a calibration scan is the most exact way of determining the MPI system function and solving the problem of MPI reconstruction. Approaches including a measurement-based MPI system function therefore outperform all approaches including modeling of either the particle behavior or the magnetic fields.

The measurement-based system function is acquired by moving a delta sample of SPIO tracer particles through the FOV and measuring the induced signal for every spatial position r_i of the sample. For a given concentration c_δ and volume V_δ of the delta sample, the induced signal in frequency space is equal to

$$\hat{u}_n(r_i) = \int_\Omega \delta(r_i) \, \hat{s}_n(r_i) \, c(r_i) \, \mathrm{d}^3 r = s_n(r_i) \, c_\delta V_\delta, \tag{2.49}$$

and the system function can be determined via

$$\hat{s}_n(r_i) = \frac{\hat{u}_n(r_i)}{c_\delta V_\delta}. \tag{2.50}$$

The major advantage of using a measurement-based system function is the fact that the sophisticated particle dynamics as well as all scanner characteristics are included.

However, the disadvantage of this approach is the highly time consuming measurement process. The acquisition of the system function takes about nine hours even for a small grid of $20 \times 20 \times 20$ image voxels [Rog11]. The reason for this is the relation between the SNR, the particle concentration c, and the measurement time T_meas

$$\text{SNR} \propto c \sqrt{T_\text{meas}}. \tag{2.51}$$

Not only a long measurement time T_{meas}, but also a high particle concentration c positively influences the SNR. Long measurements are certainly unwanted, but a high particle concentration also leads to undesirable effects, since particle-particle interactions have an increased influence for higher particle concentrations [Rog11].

2.6.4 Model-based System Function

Another approach of acquiring the MPI system function has been presented in [KBS+09] for 1D data and in [KBS+10b] for 2D data and utilizes modeling of the involved magnetic fields as well as the particle behavior. The only measured component is the transfer function of the receive chain.

If the process of determining the transfer function is considerably faster than the process of measuring the system function, this approach represents a first step in the direction of avoiding long calibration scans. Furthermore, the results presented in [KBS+09] and [KBS+10b] are of great importance regarding the physical description of the MPI imaging process. As for CT, a complete model describing MPI and enabling efficient and fast reconstruction is desired. Reconstruction using a model-based MPI system function not yet leads to an image quality comparable to that of images reconstructed with a measurement-based system function. However, a proof-of-concept for applying physical models to achieve MPI reconstruction is certainly demonstrated in [KBS+09] and [KBS+10b].

Enhancing both, modeling of the magnetic fields as well as the complicated particle dynamics will further improve image quality achieved when using model-based system function in MPI reconstruction.

2.6.5 Hybrid Approach

To find a compromise between long calibration scan times and insufficient modeling of the physics behind MPI, a hybrid approach for determining the system function was presented in [GGB+11] for 1D data.

To include the complicated particle dynamics, which are far less understood than the involved magnetic fields, the magnetization curve of an SPIO particle sample was measured using a magnetic particle spectrometer (MPS) [BKS$^+$09, Bie12]. The results were then included into the model-based approach leading to a more realistic description of the MPI signal chain.

In addition to a measured particle magnetization, also the magnetic fields could be measured and included into the model. This could be done at fewer spatial positions than used in the system function and extended by interpolation, which would decrease calibration scan time.

Building preferably exact MPI scanners, which magnetic fields can fully be determined in simulations, as well as providing a more sophisticated particle model will lead to improvements in all approaches of determining the MPI system function, which include simulations.

2.6.6 1D System Function for Ideal Conditions

In general, it is not possible to analytically calculate the MPI system function. Only in the highly simplified case of 1D imaging, perfectly homogeneous magnetic fields as well as linear gradients it is possible to show that the nth harmonic of the system function can be expressed via a convolution of the derivative of the magnetization curve with the $(n-1)$th Chebyshev polynomial of the second kind [RWGB09]

$$\hat{s}_n = -\frac{2i}{T} \int_{-A_D}^{A_D} M'(H_S - H_D) U_{n-1}(H_D/A_D) \sqrt{1 - (H_D/A_D)^2}\, dH_D \qquad (2.52)$$

$$= -\frac{2i}{T} M'(H_S) * \left(U_{n-1}(H_S/A_D) \sqrt{1 - (H_S/A_D)^2} \right). \qquad (2.53)$$

The 1D selection and drive field are denoted H_S and H_D, T is the drive field period, and A_D is the drive field amplitude. In the case of an 1D selection field $H_S = gx$, the frequency components of the MPI signal can thus be expressed via

$$\hat{s}_n = -\frac{2i}{T} M'(gx) * \left(U_{n-1}(gx/A_D) \sqrt{1 - (gx/A_D)^2} \right). \qquad (2.54)$$

This relation is only valid when neglecting relaxation of the SPIO tracer particles. As soon as relaxation arises, which will to some extend always be the case in a realistic particle system, the analytic expression derived for the system function is no longer true.

However, these results convey one important message: the MPI system function is highly structured. This finding motivates all efforts on analyzing the MPI system function to simplify MPI reconstruction such as symmetry consideration or the exploitation of sparsity.

2.7 Reconstruction

The system function will always be acquired for a discrete number of spatial positions. Therefore, a discrete representation of the MPI signal equation is needed. In this thesis, system matrix based reconstruction will always be performed in frequency space. Hence, the signal equation in frequency space given in Eq. (2.47) will be considered. The volume of interest is subdivided into K equally sized voxels of volume ΔV and Eq. (2.47) is transformed into

$$\hat{u}_n \approx \Delta V \sum_{k=1}^{K} \hat{s}_n(\mathbf{r}_k) c(\mathbf{r}_k). \tag{2.55}$$

In the case of considering a finite number N of harmonics, equation (2.55) can be written in matrix vector representation [Rog11]

$$\mathbf{u} \approx \mathbf{S}\mathbf{c}, \tag{2.56}$$

where

$$\mathbf{u} = (\hat{u}_n)_{n=1}^{N} \in \mathbb{C}^N \tag{2.57}$$

$$\mathbf{c} = (c(\mathbf{r}_k))_{k=1}^{K} \in \mathbb{R}^K$$

$$\mathbf{S} = (\Delta V \hat{s}_n(\mathbf{r}_k))_{n=1,\ldots N, k=1,\ldots K} \in \mathbb{C}^{N \times K}.$$

Acquiring a map of the SPIO particle concentration within a patient is the major goal of MPI. Solving the inverse problem

$$\mathbf{c} \approx \mathbf{S}^+ \mathbf{u}, \tag{2.58}$$

will lead to the final SPIO tracer distribution, where $\mathbf{S}^+$ denotes the pseudoinverse matrix of $\mathbf{S}$.

To solve the linear system of equations either singular value decomposition or iterative methods like conjugate gradient normal residual (CGNR) [GvL93] and Kaczmarz [Kac37] can be applied. A detailed description on efficient frequency based MPI reconstruction is provided in [Kno11].

2.8 MPI Signal Chain

To perform imaging, all of the phenomena presented so far need to work together. Furthermore, several electrical components [SGK⁺rg] are needed to generate and detect the electromagnetic fields constituting the basis for MPI. To summarize the imaging process as well as all components needed to realize the same, the MPI signal chain shall be discussed in the following.

To excite the particles and to steer the FFP or the FFL, respectively, through the FOV, a sinusoidal excitation field is needed. A sinusoidal signal is generated at the PC and transformed into an analog voltage signal using a digital-to-analog converter (DAC). Subsequently, the signal is amplified. Unfortunately, the amplification will exhibit non-linearities onto the signal, expressed by the total harmonic distortion of the amplifier. Since imaging is based on the analysis of the harmonics generated by the non-linear magnetization behavior of the SPIO tracer particles, it is of great importance to provide a purely sinusoidal excitation signal.

Hence, the amplified signal needs to be filtered subsequently using a band pass filter (BPF). A preferably pure sinusoidal signal is then applied to the transmit coil assembly, where the SPIO tracer particles will perform their characteristic non-linear magnetization change as discussed in section 2.2.4. Due to Faraday's law, a voltage, which is proportional to the change of the magnetization with respect to time, is induced in receive coils. In addition to the particle signal, which contains the base frequency of the excitation signal as well as higher harmonics, the transmit signal directly couples into the receive coil. This signal is magnitudes larger than the particle signal. To separate the particle signal from the excitation signal, filtering is needed again. This time, however, a band stop filter (BSF) is applied to damp the excitation frequency. Unfortunately, the filtering also damps the base frequency of the particles. The remaining signal only contains higher harmonics of the particle signal and a loss in information needs to be accepted. This fact influences MPI reconstruction, as discussed in section 2.7. The voltage induced due to the magnetization change of the SPIO tracer particles is very low and hence needs to be amplified using a low noise amplifier (LNA). Afterwards, it is loaded into the PC, saved, and processed on the hard disk of the PC. The MPI signal chain is illustrated in Fig. 2.14.

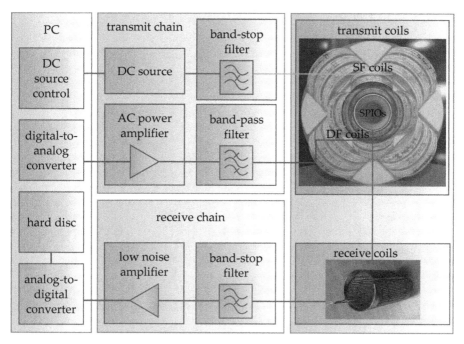

Figure 2.14 MPI signal chain for the FFL scanner presented in chapter 8. For simplicity reasons, only one spatial component is presented.

CHAPTER 3

Introduction of a Field Free Line for Magnetic Particle Imaging

Sensitive spot imaging methods in general hold an intrinsic disadvantage: an increase in spatial resolution inevitably leads to a decrease in signal to noise ratio (SNR). MPI applying an FFP for spatial encoding unfortunately is no exception. However, there is a solution to this problem, because the nature of the MPI imaging process allows for different shapes of the selection field. Implementing appropriate field generating components, it is possible to achieve a selection field featuring an FFL. The simultaneous encoding scheme applied in FFL imaging will considerably increase the sensitivity of MPI and provides a solution for the counteracting of resolution and SNR in the case of the sensitive spot FFP method.

In this chapter a motivation for the use of an FFL replacing the conventionally used FFP characterizing the MPI selection field is provided. Not only the increase in sensitivity will be discussed in this regard, but also the possibility to use efficient Radon-based reconstruction algorithms arising for a line detection scheme. In a final step, the great potential of the FFL method concerning image quality via an increase in sensitivity and SNR is demonstrated in a simulation study.

The requirements on the magnetic fields and the FFL trajectory needed for dynamic FFL imaging in MPI are described. Magnetic field generation is closely related to the design of the field generating components. For that reason, the

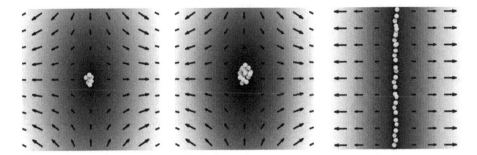

Figure 3.1 Illustration of three different MPI selection fields in arbitrary units. The right and center image show FFP selection fields, where the field gradient of the left image equals twice the field gradient of the center image. Hence, the resolution achieved with the selection field on the left will be superior, while a higher SNR will be achieved with the selection field in the center. The right image shows an FFL selection field, which combines a high gradient and spatial resolution with a high SNR.

coil assembly needed for the implementation of an FFL scanner is examined and a chronological outline of the development of FFL scanner designs is presented. Following that, different FFL imaging techniques utilizing either a static or a dynamic FFL selection field are described and discussed.

3.1 Motivation

The motivation for using an FFL shaped selection field for MPI involves two major arguments. The first is the increase in sensitivity. The second is the emergence of efficient Radon-based reconstruction algorithms, which will considerably reduce reconstruction time. Both of these arguments will be discussed in detail in the following sections.

3.1.1 Increased Sensitivity

Conventional MPI uses an FFP to scan the region of interest [GW05, WGR+09]. The resolution as well as the SNR of this sensitive spot imaging method depend on the gradient strength of the FFP selection field. To understand why spatial

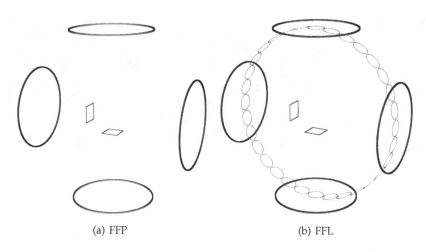

(a) FFP   (b) FFL

Figure 3.2 FFP and FFL scanner used for the presented simulation study.

resolution and SNR are counteracting in the FFP method, the influence of the gradient strength on these parameters needs to be considered.

The higher the field gradient, the smaller is the resulting region of unsaturated SPIO tracer particles contributing to the acquired MPI signal. And the smaller the resolved region, the higher is the spatial resolution.

Unfortunately, the decrease of the resolved region simultaneously leads to a reduction of the amount of signal generating SPIO tracer particles and in turn to lower SNR. This is illustrated in Fig. 3.1, where in the left and center image unsaturated SPIO particles located within FFP fields of two different gradient strengths are illustrated. The gradient of the FFP field on the left is twice as high as the gradient of the FFP field in the center image. Hence, the region, in which SPIO tracer particles are unsaturated is larger, and more SPIO tracer particles are detected at a time.

In conclusion, in the case of FFP imaging, an increase in gradient strength does improve the spatial resolution but inevitably lowers the SNR at the same time.

This is an inherent problem concerning MPI when using a sensitive spot spatial encoding method. A way to overcome this problem is to use a simultaneous encoding scheme applied by using the signal generated by tracer material located along a line. This way, a higher amount of SPIO tracer particles responds

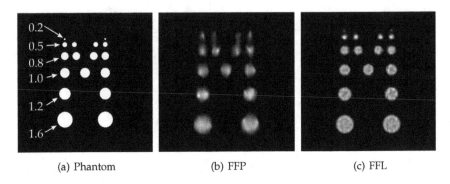

(a) Phantom   (b) FFP   (c) FFL

Figure 3.3 MPI images simulated and reconstructed with identical parameters. (a) shows the phantom used for simulation, where the diameter of the dots are given in mm. (b) shows the image simulated with an FFP scanner and in (c) the image simulated with an FFL scanner is illustrated. The gain in sensitivity and hence image quality using the FFL encoding scheme is clearly visible.

to the excitation field and hence, the SNR is increased, while retaining a constant gradient strength. This is visualized in the right image of Fig. 3.1. This method was first introduced in 2008 by Weizenecker and Gleich in [WGB08], where the great potential of FFL imaging in terms of increasing the sensitivity was demonstrated in a simulation study.

3.1.2 FFL versus FFP Imaging: a Simulation Study

In the presented MPI simulation study, which is similar to the one presented in [WGB08], reconstructed images obtained using the FFP and the FFL encoding scheme are compared. The selection field gradient strength of both scanner designs was chosen to be 2.5 $Tm^{-1}\mu_0^{-1}$. The scanners used for the simulation study are identical to the ones used in [WGB08], which resemble an almost ideal scanner regarding the relation of the scanners' bore diameter of 1 m compared to the field of view (FOV) of $14 \times 14\,mm^2$. However, the electrical power consumption of these scanners exceeds the limits for technical feasibility. These designs rather constitute a theoretical proof-of-principle for the imaging method itself and they are the origin for a development of scanners resulting in an efficient design enabling imaging experiments.

The FFP scanner used in the simulation study is shown in Fig. 3.2(a). It consists of one combined Maxwell Helmholtz coil pair orientated along the x-axis generating the FFP selection field as well as the drive field in x-direction. Another Helmholtz coil pair generates the drive field in y-direction. Hence, due to Maxwell's equations a higher resolution in x-direction is expected, since the FFP has an elliptical shape and the selection field features a higher field gradient in x-direction.

The FFL scanner design (Fig. 3.2(b)) consists of 16 Maxwell coil pairs generating the selection field. Two separate Helmholtz coil pairs are responsible for drive field generation. In both cases the drive field amplitude is set to $20\,\mathrm{mT}\mu_0^{-1}$ and the drive field frequency to 25.25 kHz according to [WGB08].

Performing an MPI simulation study, the astonishing results shown in Fig. 3.3 were achieved. The reconstructed image using the FFP encoding scheme is illustrated in Fig. 3.3(b), while Fig. 3.3(c) shows the image obtained using the FFL encoding scheme. The underlying phantoms with the dimensions given in mm are visible in Fig. 3.3(a). The increase in spatial resolution for the FFL images compared to the FFP images is clearly visible. But not only the overall spatial resolution is superior in the FFL images, but another important advantage can be observed: the homogeneity of the spatial resolution.

As discussed in the previous chapter 2, the non-existence of magnetic monopoles expressed in Maxwell's equation $\nabla B = 0$ leads to an elliptically shaped FFP, since the gradient will always be higher and conversely orientated in one spatial direction. The FFP selection field used for the simulation of Fig. 3.3(b) has its highest gradient strength in horizontal direction. Consequently, higher spatial resolution in this direction can be observed. Analyzing Fig. 3.3(b), the 0.5 mm and 0.8 mm dots are smeared in vertical direction, while in horizontal direction they are resolved. The 0.2 mm dot is not resolved. In contrast to the FFP image, the spatial resolution of the FFL image is not only higher, but also spatially homogeneous. The 0.8 mm and the 0.5 mm dots are separated in both directions. Even the 0.2 mm dot is clearly resolved. These results emphasize the potential held by the FFL imaging method to considerably increase sensitivity and spatial resolution of MPI.

Regarding these promising results, the question arises why the first MPI scanners [GW05, SKB+09, GC11] were using the FFP encoding scheme. To answer this question FFL field generation as well as the design of the corresponding

MPI scanners need to be discussed.

The encoding scheme introduced in [WGB08] not only demands for the generation, but also for the rotation and translation of the FFL field. The setup of electromagnetic coils and permanent magnets needed for magnetic field generation in FFL imaging was initially assumed to be far too complicated and inefficient with respect to electrical power consumption to be feasible for technical implementation. Fortunately, this skepticism diminished due to essential improvements in the design of dynamic FFL imagers [KSBB10, KEB+10, ESKB12], which are, to some extent, also part of this thesis. A chronological outline of FFL scanner designs will be provided in section 3.3.

A considerable increase in sensitivity is not the only advantage regarding a line encoding scheme. The possibility to use well-known and powerful Radon-based reconstruction algorithms provides additional motivation to use the FFL method and will be discussed in the subsequent section.

3.1.3 Efficient Radon-based Reconstruction Algorithms

The improvements of the efficiency of the FFL imager designs encouraging the vision of imaging experiments also motivated further considerations on the reconstruction process. Since the simultaneous encoding scheme applied in the FFL imaging method uses the signal of SPIO tracer particles located along a line, analogies to first generation CT may be utilized. It was shown in [KES+11] that the transformation of the FFL data into Radon-space opens up the possibility to use well-known and powerful reconstruction algorithms considerably decreasing reconstruction time.

The mathematical derivation of this transformation as well as reconstructed results and a simulation study on the limitations of this method is presented in chapter 7. Furthermore, the influence of scanner optimization on reconstructed results is discussed.

The conditions on the FFL field generation resulting in the correct FFL trajectory enabling conventional as well as Radon-based reconstruction are described in the following section.

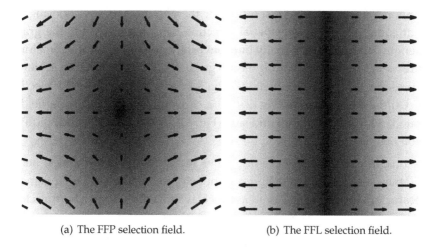

(a) The FFP selection field. (b) The FFL selection field.

Figure 3.4 Comparison of FFP and FFL selection field.

3.2 The FFL Field and Trajectory

Spatial encoding in MPI is achieved via the use of magnetic gradient fields, the selection fields. In section 2.5, this was described for the conventional FFP imaging method. FFL imaging uses a similar method involving gradient fields, however, two major facts differentiate the FFL from the FFP selection field.

The first very obvious difference lies in the shape of the field free region, i.e. the region in which SPIO tracer particles contribute to the acquired signal. The FFL field, as the name suggests, is characterized by a line of zero field strength and a constant gradient perpendicular to its alignment. Mathematically a magnetic field featuring an FFL oriented along the x-axis with a constant field gradient $g^{FFL,x}$ is described by

$$H^{FFL,x}(r) = g^{FFL,x} \begin{pmatrix} 0 & 0 & 0 \\ 0 & 1 & 0 \\ 0 & 0 & -1 \end{pmatrix} r$$

$$= g^{FFL,x} \, \mathbf{G}^{FFL,x} \, r. \tag{3.1}$$

$\mathbf{G}^{FFL,x}$ is the gradient matrix of the x-orientated FFL field. Via the shape of this matrix it is obvious, why Gauss's law for magnetism $\nabla \cdot \mathbf{B} = 0$ does not lead to an inhomogeneous spatial resolution for the FFL method. By choosing the gradient equal to zero in one spatial direction, i.e. along the FFL, the gradient strength can be chosen equally but conversely orientated for the remaining two directions. This way, a homogeneous gradient distribution can be achieved while preserving Maxwell's equations. Fig. 3.4 compares the two MPI selection fields. On the left hand side (Fig. 3.4(a)) a conventional FFP selection field is illustrated, while on the right hand side (Fig. 3.4(b)) the newly introduced FFL selection field is shown.

The second important difference originates in the FFL field dynamics, hence the FFL trajectory. For FFP imaging, a static selection field provides a magnetic field featuring an FFP in the center, which is moved on a specific trajectory via additional drive fields as described in sections 2.5 and 2.4. Since they are of purely static nature, it is possible to generate the FFP selection field using permanent magnets rather than resistive coils. This considerably reduces the total power consumption of the corresponding MPI scanner and hence lowers implementation effort.

The situation is different for FFL imaging, however. To make use of FFL imaging in MPI specific conditions on the scanned trajectory have to be fulfilled. The reconstruction of the collected line integrals of the particle concentration along the FFL requires a rotation of at least 180 degrees as well as translation of the FFL [Rad17]. Initiated by analogies to the acquisition scheme used in first generation CT, the slowly rotated FFL exhibits a fast translation.

As a result, the selection field of a dynamic FFL imaging device is no longer static, but rotates with a certain frequency f_S, which is much lower than the excitation frequency of the drive field f_D, and hence, the FFL translation. The frequency of the selection field amounts to some 100 Hz, while the drive field frequency is set to a value around 25 kHz similar to FFP imaging. In addition to the FFL field translation, the drive field is also responsible for the excitation of the SPIO tracer particles. These combined dynamics result in a radial FFL trajectory as illustrated in Fig. 3.5.

The important aspect regarding the interaction between FFL selection and drive field is that the drive field needs to be orientated perpendicular to the FFL alignment at any time. Selection field generation providing an FFL field with a complete 360 degree rotation will be discussed in detail in chapter 4. Considering a given FFL selection field and the orientation of the drive field perpendicular to

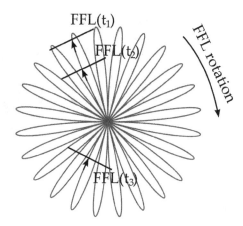

Figure 3.5 Radial FFL trajectory. The FFL alignment is illustrated for three different time points t_1, t_2, and t_3.

the FFL alignment, a radial 2D FFL trajectory can be achieved by two Helmoltz coil pairs generating the drive field, which need to carry the following currents [KBS$^+$09]

$$I_{A,x}(t) = I_0 \sin(2\pi f_D t) \cos(2\pi f_S t) \tag{3.2}$$

and

$$I_{A,y}(t) = I_0 \sin(2\pi f_D t) \sin(2\pi f_S t). \tag{3.3}$$

The scanning density of the radial FFL trajectory is determined by the frequency ratio N_p

$$N_p = \frac{f_D}{f_S}. \tag{3.4}$$

Since magnetic field generation for FFL imaging in MPI holds novel challenges, the geometry of the MPI scanning device is sophisticated. The evolution of the FFL scanner designs is discussed in the subsequent section.

3.3 The FFL Scanner Design

Although simulation results on the sensitivity promise great potential, the method of FFL imaging in MPI was doubted even by its inventors, due to the fact that

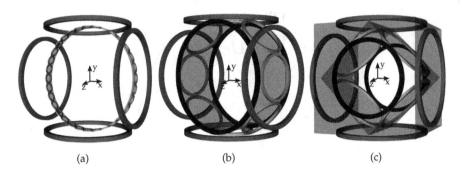

(a)   (b)   (c)

Figure 3.6 Evolution of the proposed FFL scanner design. Setup (a) illustrates the first scanner setup consisting of 16 Maxwell coil pairs and two larger surrounding Helmholtz coil pairs. In setup (b) the number of coil pairs on the circle has been reduced to 4 and the z-coils are introduced. Setup (c) shows a sketch of the implemented coil design.

the first FFL imager proposed in [WGB08] was rather inefficient regarding its coil design. The first proposed scanner setup capable of generating an FFL field complying with the introduced requirements, consisted of 16 small Maxwell coil pairs located at equidistant angles on a circle and 2 larger surrounding Helmholtz coil pairs forming a cube. A sketch of the setup is illustrated in Fig. 3.6(a). Applying appropriate currents to the ring of coils allows for the generation of the FFL as well as an arbitrary rotation in the xy-plane. Hence, the circle of coils is responsible for selection field generation. The FFL drive field imposing a translation is realized via the cube of Helmholtz coil pairs. Unfortunately, this scanner setup would consume more than 1000 times the power of an FFP scanner of equal size and gradient strength [KEB+10] and must therefore be considered unfeasible for practical implementation. However, the prospect of a substantial improvement in sensitivity motivated further effort in the field of FFL imaging in MPI.

Within the scope of further investigations it was mathematically proven that 3 Maxwell coil pairs located on a circle at equidistant angles suffice to generate and rotate the FFL field [KSBB10]. Taking the field quality into account, however, the choice of 4 Maxwell coil pairs proved to be optimal. These findings constitute a major step towards feasibility of the FFL concept, since the reduction of scanner components results in a substantial decrease in electrical power

Figure 3.7 Introduction of the z-coils responsible for generation of the static part of the FFL selection field. As indicated by black arrows the currents in the neighboring parts of the coils cancel, while the currents in the outer parts are identical to those in a Maxwell coil pair in z-direction.

consumption. An FFL scanner consisting of a ring of 4 Maxwell coil pairs would only consume 6.9 times the power of a conventional FFP scanner [KEB+10]. Electrical power consumption actually comparable to that of an FFP scanner, however, was not attained until optimization considerations initiated the separation of the alternating and the static part of the applied selection field currents. This modification in the setup led to an overall electrical power consumption of merely $P_{FFL} = 1.6\ P_{FFP}$ [KEB+10]. In Fig. 3.7 the setup of annular arranged coils shown in Fig. 3.6(b) is approximated by rectangular coils. The static currents within the parts of neighboring coils facing each other are vice versa orientated and therefore cancel. These currents are redundant, since they do not contribute to the generated fields but do contribute to electrical power consumption. The currents on the outer sides of the coils could more efficiently be generated by a dedicated Maxwell coil pair orientated along the z-axis, the axis through the scanner setup. By introducing these z-coils only the field generating parts of the static currents are retained and hence the power loss is reduced. The corresponding scanner device is sketched in Fig. 3.6(b).

By softening the concept of the circle of coils, it is possible to further increase the field quality. The newly introduced scanner design, illustrated in Fig. 3.6(c), separates the four Maxwell coil pairs of the former circle in two pairs each located on the surface of a cube. The inner cube is rotated by 45 degrees with respect to the outer cube. In this way, it is possible to increase the size of the individual coils. In addition, the distance between the coil pairs of the inner cube is reduced compared to the circle setup. The enhanced field quality is related to the increased size of the individual coils as well as to the decrease in their distance. Ideal fields are a theoretical approach obtained for infinite distance, radius and field strength of each coil, only. Reducing the radius of the coils for a constant distance leads to a diminished field quality. In turn, reducing the distance for a constant radius will enhance the field quality. Hence, keeping the size of the setup constant, the field quality will improve when enlarging the size of the individual coils or diminishing their distance.

Up to that point, all considerations regarding the FFL imager design were of a theoretical kind. An experimental proof for the feasibiliy of magnetic field generation for FFL imaging in MPI was not provided until an FFL field demonstrator capable of generating a full FFL trajectory while consuming power as predicted by simulation and hence comparable to the power consumption of an FFP scanner was implemented as part of this thesis and presented in [EKS+11a]. For this implementation, the design shown in Fig. 3.6(c) was chosen, since it was the most efficient design known at that time.

Further improvements of the FFL scanner design will be presented in chapter 6 as part of this thesis leading not only to further reduction of the electrical power consumption, but also to a considerable increase in magnetic field quality. This is of special importance with regard to the use of efficient Radon-based reconstruction algorithms, where deformation of the FFL field would lead to artifacts in the reconstructed images. The simulated results on improving the scanner design lead to the implementation of the first dynamic FFL imaging device, which will be presented in chapter 8.

3.4 Field Free Line Imaging Techniques

There are different techniques to realize FFL imaging in MPI. In any case, rapid translation of the FFL as well as 180° rotation of the FFL with respect to the object is required to reconstruct the object. The various techniques presented in this section realize rotation of the FFL either by mechanical movement of the object or the scanner setup, respectively, or by providing a coil assembly capable of rotating the FFL by applying appropriate currents, only.

In the FFL projection imaging technique, a field free line is generated under one specific angle using either resistive electromagnetic coils or permanent magnets combined with a rapid translation of the FFL, which is realized via additional electromagnetic coils in Helmholtz configuration. Either the scanner setup or the object is then rotated to acquire reconstructable data. The first FFL images have been produced using this technique by researchers at Berkeley [GKZ$^+$12, GKZC12, KGCZC12].

In this thesis, dynamic FFL imaging will be presented. Using this method, no mechanical movement is necessary at all. This way the real-time capability of MPI is preserved.

3.4.1 Preliminaries: Static Field Free Line Generation

To apply projection FFL imaging, a static FFL is generated by the MPI selection field. To do so, either electromagnetic coils or permanent magnets can be used.

As discussed in chapter 2, Maxwell coil pairs are used to generate static FFP gradient fields. Similar, Maxwell coil pairs can be used to generate a static FFL gradient field. As illustrated in Fig. 3.8, the superposition of two FFP gradient fields with main gradient contribution in x- and y-direction leads to the formation of a gradient field characterized by an FFL along the z-direction

$$H^{\text{FFP},x}(r) + H^{\text{FFP},y}(r) = I_x p_x \begin{pmatrix} 1 & 0 & 0 \\ 0 & -\frac{1}{2} & 0 \\ 0 & 0 & -\frac{1}{2} \end{pmatrix} r + I_y p_y \begin{pmatrix} -\frac{1}{2} & 0 & 0 \\ 0 & 1 & 0 \\ = & 0 & -\frac{1}{2} \end{pmatrix} r \qquad (3.5)$$

assuming $p_x = p_y = p$ and choosing $I_x = -I_y = I$, this is

$$H^{\text{FFP},x}(r) + H^{\text{FFP},y}(r) = \frac{3}{2} I p \begin{pmatrix} 1 & 0 & 0 \\ 0 & -1 & 0 \\ 0 & 0 & 0 \end{pmatrix} r = H^{\text{FFL},z}(r). \qquad (3.6)$$

Generating such an FFL field, the relative orientation of the FFP fields in x- and y-direction needs to be chosen properly using $I_x = -I_y$. Reorientating either the FFP field in x- or y-direction by choosing the currents in the coil pairs according to $I_x = I_y$ would lead to the generation of an FFP field in z-direction

$$H^{\text{FFP},x}(r) + H^{\text{FFP},y}(r) = I p \begin{pmatrix} 1 & 0 & 0 \\ 0 & -\frac{1}{2} & 0 \\ 0 & 0 & -\frac{1}{2} \end{pmatrix} r + I p \begin{pmatrix} -\frac{1}{2} & 0 & 0 \\ 0 & 1 & 0 \\ 0 & 0 & -\frac{1}{2} \end{pmatrix} r$$

$$= Ip \begin{pmatrix} \frac{1}{2} & 0 & 0 \\ 0 & \frac{1}{2} & 0 \\ 0 & 0 & -1 \end{pmatrix} r = H^{FFP,z}(r), \tag{3.7}$$

again assuming equal coil sensitivities $p_x = p_y = p$.

It is furthermore possible to generate an FFL field along the x- and y-direction by the same assembly of coils. Only the current needs to be chosen twice as high in the coil pair orientated perpendicular to the FFL. This will be discussed in detail in section 4.2.

The first static FFL fields with respect to MPI were generated with the FFL field demonstrator, which was implemented as part of this thesis, and presented in [KES⁺10b].

3.4.2 Field Free Line Projection Imaging

One possible way of acquiring FFL data for the use in MPI with a static FFL selection field is the FFL projection imaging technique. Here, a static FFL in z-direction is generated using two Maxwell coil pairs in x- and y-direction, as described in the previous section 3.4.1. The corresponding scanner setup capable of generating the required fields is illustrated in Fig. 3.9. As the patient tube is orientated along the x-axis, 2D images of the object in the yz-plane will be acquired. To acquire imaging data, rapid FFL movement orthogonal to its alignment is required. Therefore, the FFL field is translated in y-direction, which is realized by superimposing alternating, equally orientated currents in the coil pair in y-direction as indicated by the black arrows in Fig. 3.9. This coil pair is hence used in combined Maxwell and Helmholtz mode. In FFL slice imaging, the required 180 degrees FFL rotation is achieved via a rotation of the object around the x-axis, or respectively via rotation of the scanner setup.

To further improve the applicability of this method with respect to clinical use, it is possible to limit the required rotation to 90 degrees. This is realized via a slight modification of the scanner setup by adding a coil pair in z-direction. With a combination of this additional coil pair and the existing coil pair in x-direction it is possible to generate a static FFL in y-direction. This FFL can then be translated in z-direction by superimposing alternating, equally orientated currents in the newly introduced coil pair in this direction. Using these two FFL fields

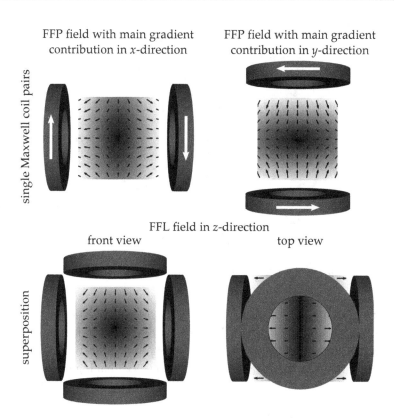

Figure 3.8 Static FFL generation with two Maxwell coil pairs. With a combination of FFP fields in x- and y-direction, it is possible to generate an FFL field along the z-axis.

along the z- and y-direction, it is possible to acquire reconstructable data with only 90 degrees rotation of the object or the scanner, respectably. The method is illustrated in Fig. 3.10, where the object is rotated by 90 degrees and at each angle the FFL is subsequently generated in z- and y-direction and translated perpendicular to its alignment.

One positive aspect of FFL slice imaging is the possibility to generate the static FFL field using permanent magnets. Without the need to use resistive coils, a considerable reduction in overall electrical power consumption of the scanner setup can be achieved. However, the need of object or scanner rotation induces collateral implementation effort, since in the case of clinical application

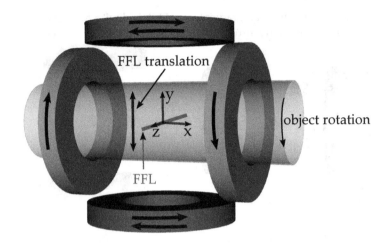

Figure 3.9 FFL projection scanner based on [KB12].

the scanner and not the object would need to be rotated as done in CT. Since one major advantage differentiating MPI from existing medical imaging technologies is its real-time capability, which will not be provided using any mechanical movement. This is a major drawback of the FFL slice imaging method.

3.4.3 Dynamic Field Free Line Imaging

A more sophisticated imaging technique is presented in this thesis, i.e. dynamic FFL imaging. In contrast to FFL projection imaging, where a static FFL is applied, not only dynamic FFL translation but also rotation is realized via applying appropriate drive and selection field currents, only. In the following, only dynamic FFL imaging will be discussed. A detailed description of the magnetic fields needed for dynamic FFL imaging as well as the connection to the field generating electromagnetic coils or permanent magnets, respectively, is provided in chapter 4.

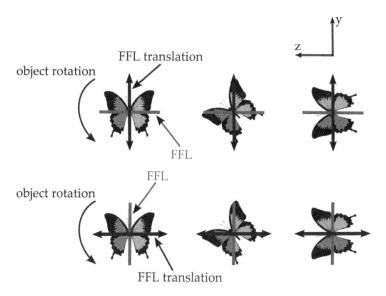

Figure 3.10 FFL slice imaging with a 90 degrees object rotation and two FFLs aligned in z- and y-direction.

Theory of Magnetic Field Free Line Generation

This chapter provides a detailed description of the theoretical foundations of magnetic field generation necessary for FFL imaging in MPI. The magnetic fields as well as the electromagnetic coils generating the same are discussed and analyzed. Taking into account the conditions on the magnetic fields mentioned in the previous chapter, i.e. generation, rotation, and translation of the FFL field, the scanner setup first introduced in [KEB+10] and shown in Fig. 4.1 will be motivated from a purely mathematical point of view. This setup was also chosen for the implementation of the first FFL field demonstrator, which will be introduced in chapter 5 .

The theoretical foundations provided in this chapter as well as the description of the FFL field demonstrator presented in chapter 5 are published in [EKS+11a] by the author of this thesis et. al. and this and the subsequent chapter 5 are based on the named publication.

4.1 FFP Field Generation

In the presented FFL scanner setup in Fig. 4.1, which is to be mathematically motivated in this chapter, all magnetic selection field components are generated via a combination of individual Maxwell constant gradient coil pairs. Therefore, the point of origin for FFL considerations will be the detailed discussion of the

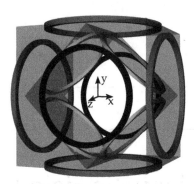

Figure 4.1 Efficient FFL scanner design mathematically motivated in this chapter and implemented as FFL field demonstrator.

fields generated by such coil pairs. These coil pairs are characterized by a uniform radius of both coils as well as by opposing currents.

Note, that the equations derived in this chapter hold for ideal coils with infinite diameter and distance. Certainly, this is not fulfilled in reality. The theoretical results of this chapter are only valid in a small area in the center of the coil assembly. Outside this area a loss in field quality has to be accepted.

The magnetic fields generated by Maxwell coil pairs in x-, y- and z-direction are given by

$$H^{\mathrm{FFP},x}(r) = I_x p_x \begin{pmatrix} 1 & 0 & 0 \\ 0 & -\frac{1}{2} & 0 \\ 0 & 0 & -\frac{1}{2} \end{pmatrix} r = I_x p_x \, \mathbf{G}^{\mathrm{FFP},x} \, r \tag{4.1}$$

$$H^{\mathrm{FFP},y}(r) = I_y p_y \begin{pmatrix} -\frac{1}{2} & 0 & 0 \\ 0 & 1 & 0 \\ 0 & 0 & -\frac{1}{2} \end{pmatrix} r = I_y p_y \, \mathbf{G}^{\mathrm{FFP},y} \, r \tag{4.2}$$

$$H^{\mathrm{FFP},z}(r) = I_z p_z \begin{pmatrix} -\frac{1}{2} & 0 & 0 \\ 0 & -\frac{1}{2} & 0 \\ 0 & 0 & 1 \end{pmatrix} r = I_z p_z \, \mathbf{G}^{\mathrm{FFP},z} \, r. \tag{4.3}$$

The coil sensitivities p_x, p_y, and p_z are determined by the properties of the corresponding coils and their distance to the center of the setup and are defined in Eq. (2.29). I_x, I_y, and I_z are the currents applied to the coil pairs. The field generated by a Maxwell coil pair is characterized by a FFP in the center of the coils and a constant vector gradient or Jacobian matrix, respectively:

$$
J_{\mathbf{H}^{FFP,i}} = \begin{pmatrix} \dfrac{\partial \mathbf{H}_x^{FFP,i}}{\partial x} & \dfrac{\partial \mathbf{H}_x^{FFP,i}}{\partial y} & \dfrac{\partial \mathbf{H}_x^{FFP,i}}{\partial z} \\[2mm] \dfrac{\partial \mathbf{H}_y^{FFP,i}}{\partial x} & \dfrac{\partial \mathbf{H}_y^{FFP,i}}{\partial y} & \dfrac{\partial \mathbf{H}_y^{FFP,i}}{\partial z} \\[2mm] \dfrac{\partial \mathbf{H}_z^{FFP,i}}{\partial x} & \dfrac{\partial \mathbf{H}_z^{FFP,i}}{\partial y} & \dfrac{\partial \mathbf{H}_z^{FFP,i}}{\partial z} \end{pmatrix}
$$

$$
= I_i \, p_i \, \mathbf{G}^{FFP,i}
$$

$$
= g^{FFP,i} \, \mathbf{G}^{FFP,i}, \quad \text{with } i = x, y, z. \tag{4.4}
$$

$G^{FFP,i}$ is normalized in such a way, that its highest value amounts to 1 and it will therefore be denoted normalized gradient matrix, while $g^{FFP,i} = I_i \, p_i$ is called the gradient of the FFP field. The gradient is identical to the maximum slope of the field. It is reached in the direction, along which the coil pair is orientated. The slope then linearily decreases until in the remaining two spatial directions it has reached its minimum of $-\frac{1}{2} g^{FFP,i}$. The gradient in direction of the coil axis is twice as high as the gradient in the remaining two directions and conversely orientated due to Maxwell's equation

$$
0 = \nabla \cdot \mathbf{B} \tag{4.5a}
$$

$$
= \mu \left(\frac{\partial H_x}{\partial x} + \frac{\partial H_y}{\partial y} + \frac{\partial H_z}{\partial z} \right), \tag{4.5b}
$$

resulting in an elliptically shaped FFP, as illustrated in Fig. 4.2(a) and 4.2(b). The fields generated by Maxwell coil pairs in x-, y-, and z-direction provided in Eq. (4.1) are linearly dependent. It is therefore possible, to generate an FFP field with main gradient contribution in z-direction by means of a linear combination of the fields generated by Maxwell coil pairs in x- and y-direction, for an appropriate choice of currents as shown in Fig. 4.2(c). The two coil pairs are assumed to be equally manufactured and located with respect to the center. Therefore, $p_x = p_y =: p$ is assumed. For the generated magnetic field, the following relation is obtained

$$
\mathbf{H}^{FFP,z}(r) \stackrel{!}{=} \mathbf{H}^{FFP,x}(r) + \mathbf{H}^{FFP,y}(r)
$$

$$
= I_x \, p \, \mathbf{G}^{FFP,x} \, r + I_y \, p \, \mathbf{G}^{FFP,y} \, r
$$

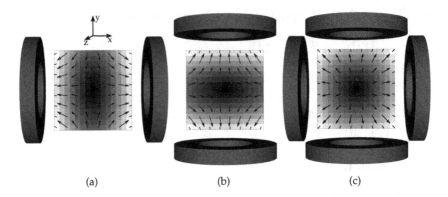

(a)   (b)   (c)

Figure 4.2 Assemblies of Maxwell coil pairs generating FFP fields. Setup (a) shows a Maxwell coil pair orientated along the x-axis. The gradient is twice as high in the direction of the coil pair. Setup (b) illustrates the same setup in y-direction. In setup (c) the superposition of setup (a) and (b) leads to an FFP field with main gradient contribution in z-direction and, therefore, a symmetric field in the xy-plane.

$$= p\left(I_x - \tfrac{1}{2}I_y, \quad -\tfrac{1}{2}I_x + I_y, \quad -\tfrac{1}{2}I_x - \tfrac{1}{2}I_y\right) r. \tag{4.6}$$

By comparing Eq. (4.6) to Eq. (4.3), the following condition on the currents emerges

$$I_x = I_y = -I_z. \tag{4.7}$$

Equivalently, an FFP field in x- or y-direction, respectively, can be generated by superimposing the fields of Maxwell coil pairs orientated along the remaining two spatial axes. Hence, two orthogonal Maxwell coil pairs along two spatial axes are always sufficient to generate FFP fields with main gradient contribution along any of the three directions in space.

4.2 On Axis FFL Field Generation

In the presented thesis, the method of applying an FFL selection field is discussed. The generation of a magnetic field characterized by a static FFL along one of the three spatial directions is therefore introduced in the following. The question to be answered is, how to generate an FFL field most efficiently from the fields provided by an assembly of Maxwell coil pairs. As a first step, a field characterized by an FFL orientated along the x-axis is examined

$$H^{\text{FFL},x}(r) = g^{\text{FFL},x}\begin{pmatrix} 0 & 0 & 0 \\ 0 & 1 & 0 \\ 0 & 0 & -1 \end{pmatrix} r$$

$$= g^{\text{FFL},x}\, G^{\text{FFL},x}\, r. \tag{4.8}$$

The factor $g^{\text{FFL},x}$ is the gradient of the generated FFL field. It is, as for the FFP field, identical to the highest slope, which is for Eq. (4.8) reached perpendicular to the FFL in y-direction. Then the slope linearly decreases until the minimum slope of $-g^{\text{FFL},x}$ is reached in z-direction. The gradient is conditioned by the properties of the generating coils and the geometry of the coil assembly as well as by the applied currents.

Ideally, the desired FFL field is generated by a preferably simple assembly of Maxwell coil pairs. An FFL field orientated in x-direction shall be obtained by a linear combination of the fields generated by two opposing current coil pairs orientated along the x- and y- axis, as illustrated in Fig. 4.3(a). Again, the coils are assumed to be equally manufactured and located with respect to the center. Therefore, the sensitivities are set to $p_x = p_y =: p$. The currents may be varied, to obtain the desired magnetic field

$$H^{\text{FFL},x}(r) \overset{!}{=} I_x p\, G^{\text{FFP},x}(r) + I_y p\, G^{\text{FFP},y}(r)$$

$$= I_x p\begin{pmatrix} 1 & 0 & 0 \\ 0 & -\frac{1}{2} & 0 \\ 0 & 0 & -\frac{1}{2} \end{pmatrix} r + I_y p\begin{pmatrix} -\frac{1}{2} & 0 & 0 \\ 0 & 1 & 0 \\ 0 & 0 & -\frac{1}{2} \end{pmatrix} r$$

$$= p\begin{pmatrix} I_x - \frac{1}{2}I_y & 0 & 0 \\ 0 & -\frac{1}{2}I_x + I_y & 0 \\ 0 & 0 & -\frac{1}{2}I_x - \frac{1}{2}I_y \end{pmatrix} r. \tag{4.9}$$

Comparison with Eq. (4.8) leads to the following system of linear equations

$$p\left(I_x - \tfrac{1}{2}I_y\right) = 0$$
$$p\left(-\tfrac{1}{2}I_x + I_y\right) = g^{\text{FFL},x} \tag{4.10}$$
$$p\left(-\tfrac{1}{2}I_x - \tfrac{1}{2}I_y\right) = -g^{\text{FFL},x}.$$

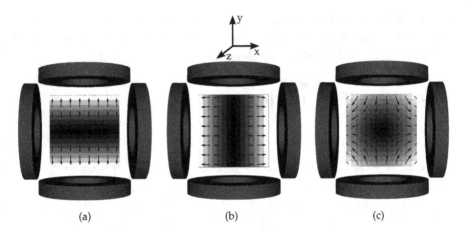

(a)   (b)   (c)

Figure 4.3 Assemblies of Maxwell coil pairs generating static FFL fields. Two Maxwell coil pairs along the x- and y-direction are sufficient to generate a magnetic field characterized by an FFL in x- (setup (a)), y- (setup (b)), and z-direction (setup (c)).

For a particular gradient $g^{\text{FFL},x}$, and in case of a given sensitivity p, solving Eq. (4.10) results in the following choice of currents

$$I_x = \frac{2\,g^{\text{FFL},x}}{3p} \quad \text{and} \quad I_y = \frac{4\,g^{\text{FFL},x}}{3p}. \tag{4.11}$$

Complying with the currents in Eq. 4.11, it is possible to generate a constant gradient field characterized by an FFL along the x-direction, using merely two Maxwell coil pairs. Equivalently, it is possible to generate an FFL field along the y-direction by interchanging the current values I_x and I_y in the corresponding coils. As already verified, it is possible to emulate the field generated by a Maxwell coil pair in z-direction via the fields of Maxwell coil pairs in x- and y-direction. It should therefore be possible to generate a field with an FFL orientated along the z-axis by the two introduced Maxwell coil pairs in x- and y-direction

$$H^{\text{FFL},z} = g^{\text{FFL},z} \begin{pmatrix} 1 & 0 & 0 \\ 0 & -1 & 0 \\ 0 & 0 & 0 \end{pmatrix} r$$

$$\overset{!}{=} H^{\text{FFL},x} + H^{\text{FFL},y}$$

$$= I_x p \begin{pmatrix} 1 & 0 & 0 \\ 0 & -\frac{1}{2} & 0 \\ 0 & 0 & -\frac{1}{2} \end{pmatrix} r + I_y p \begin{pmatrix} -\frac{1}{2} & 0 & 0 \\ 0 & 1 & 0 \\ 0 & 0 & -\frac{1}{2} \end{pmatrix} r. \tag{4.12}$$

This can be solved analogously to Eq. (4.10), leading to

$$I_x = \frac{2\, g^{\mathrm{FFL},z}}{3p} \quad \text{and} \quad I_y = -\frac{2\, g^{\mathrm{FFL},z}}{3p}, \tag{4.13}$$

as introduced in section 3.4.1. At this point, it has been shown that a static FFL in all three spatial directions can be generated by only two orthogonal Maxwell coil pairs.

4.3 FFL Field Rotation

The challenge to be coped with is to generate an arbitrarily rotated FFL field in the xy-plane, while keeping the coil setup static in space. Starting from an FFL field orientated along the x-axis (Eq. (4.8)) and applying a counterclockwise rotation by an arbitrary angle θ, the field providing an FFL along θ is given by

$$H^{\mathrm{FFL},\theta}(r) = \mathbf{R}^{\theta} g^{\mathrm{FFL},x} \mathbf{G}^{\mathrm{FFL},x} \mathbf{R}^{-\theta} r$$

$$= g^{\mathrm{FFL},x} \begin{pmatrix} \cos\theta & -\sin\theta & 0 \\ \sin\theta & \cos\theta & 0 \\ 0 & 0 & 1 \end{pmatrix} \begin{pmatrix} 0 & 0 & 0 \\ 0 & 1 & 0 \\ 0 & 0 & -1 \end{pmatrix} \begin{pmatrix} \cos\theta & \sin\theta & 0 \\ -\sin\theta & \cos\theta & 0 \\ 0 & 0 & 1 \end{pmatrix} r$$

$$= g^{\mathrm{FFL},x} \begin{pmatrix} \cos\theta & -\sin\theta & 0 \\ \sin\theta & \cos\theta & 0 \\ 0 & 0 & 1 \end{pmatrix} \begin{pmatrix} 0 & 0 & 0 \\ -\sin\theta & \cos\theta & 0 \\ 0 & 0 & -1 \end{pmatrix} r$$

$$= g^{\mathrm{FFL},x} \begin{pmatrix} \sin^2\theta & -\sin\theta\cos\theta & 0 \\ -\sin\theta\cos\theta & \cos^2\theta & 0 \\ 0 & 0 & -1 \end{pmatrix} r$$

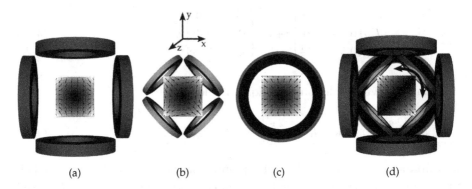

Figure 4.4 The superposition of the fields $\mathbf{H}^{a,\theta}$, generated by setup (a), $\mathbf{H}^{b,\theta}$, generated by setup (b), and $\mathbf{H}^c$, generated by setup (c), provides an rotating FFL field. Hence, the combination of these three components adds up to setup (d).

$$= g^{\mathrm{FFL},x} \begin{pmatrix} \frac{1}{2} - \frac{1}{2}\cos 2\theta & -\frac{1}{2}\sin 2\theta & 0 \\ -\frac{1}{2}\sin 2\theta & +\frac{1}{2} + \frac{1}{2}\cos 2\theta & 0 \\ 0 & 0 & 1 \end{pmatrix} r$$

$$= g\, \mathbf{G}^{\mathrm{FFL},\theta} r. \tag{4.14}$$

The scalar $g = g^{\mathrm{FFL},x}$ denotes the gradient of the rotating FFL field and $\mathbf{R}^\theta$ is the rotation matrix for a rotation in the xy-plane. The normalized gradient matrix $\mathbf{G}^{\mathrm{FFL}}_\theta$ is not diagonal for $\theta \neq n\frac{\pi}{2}$, $n \in \mathbb{N}$ and therefore it cannot be generated by a linear combination of $\mathbf{H}^{\mathrm{FFP},x}$, $\mathbf{H}^{\mathrm{FFP},y}$, and $\mathbf{H}^{\mathrm{FFP},z}$. Hence, a combination of Maxwell coil pairs along the x-, y-, and z-direction will not suffice to generate a rotating FFL field.

Taking a closer look at the normalized gradient matrix of the rotated FFL field (Eq. (4.14)), however, will lead to a very intuitive motivation for the coil setup introduced in Fig. 4.1, which will also be experimentally verified in the course of this thesis.

The FFL field $\mathbf{H}^{\mathrm{FFL},\theta}$ can be subdivided in three parts

$$\mathbf{H}^{\mathrm{FFL},\theta}(r) = g\left(\mathbf{H}^{a,\theta}(r) + \mathbf{H}^{b,\theta}(r) + \mathbf{H}^c(r)\right), \tag{4.15}$$

where

$$H^{a,\theta} = \begin{pmatrix} -\frac{1}{2}\cos 2\theta & 0 & 0 \\ 0 & \frac{1}{2}\cos 2\theta & 0 \\ 0 & 0 & 0 \end{pmatrix} r = \frac{1}{2}\cos 2\theta\, \mathbf{G}^a r \qquad (4.16)$$

$$H^{b,\theta} = \begin{pmatrix} 0 & -\frac{1}{2}\sin 2\theta & 0 \\ -\frac{1}{2}\sin 2\theta & 0 & 0 \\ 0 & 0 & 0 \end{pmatrix} r = -\frac{1}{2}\sin 2\theta\, \mathbf{G}^b r \qquad (4.17)$$

$$H^c = \begin{pmatrix} \frac{1}{2} & 0 & 0 \\ 0 & \frac{1}{2} & 0 \\ 0 & 0 & -1 \end{pmatrix} r = \mathbf{G}^c r\, . \qquad (4.18)$$

Obviously, $H^{a,\theta}$ is an FFL field aligned along the z-axis, with $g_a^{\mathrm{FFL},z} = \frac{1}{2}\cos 2\theta$ and $\mathbf{G}^a = \mathbf{G}^{\mathrm{FFL},z}$. As proven in the previous section, such a field can be generated by two Maxwell coil pairs orientated along the x- and y-axis, as illustrated in Fig. 4.4(a).

$H^{b,\theta}$ does also describe an FFL field along the z-direction, rotated by 45 degrees. This can easily be verified by computing

$$R^{-\frac{\pi}{4}}\, \mathbf{G}^b\, R^{\frac{\pi}{4}} = \mathbf{G}^a = \mathbf{G}^{\mathrm{FFL},z}. \qquad (4.19)$$

The gradient of this field is $g_b^{\mathrm{FFL},z} = -\frac{1}{2}\sin 2\theta$. To generate $H^{b,\theta}$, two orthogonal Maxwell coil pairs rotated by 45 degrees around the z-axis are introduced in Fig. 4.4(b).

The last remaining component is the field H^c, which is equal to an FFP field in z-direction. This field is most efficiently generated by one Maxwell coil pair orientated along the z-direction, shown in Fig. 4.4(c). It is a static field, since it does not depend on the angle of the FFL. The z-coils, introduced in section 3.3 for phenomenological reasons, are substantiated with a theoretical motivation now.

The superposition of the fields $H^{a,\theta}$, $H^{b,\theta}$, and H^c results in an FFL field rotated by the angle θ in the xy-plane. The combination of the three generating assemblies leads to the final scanner setup, illustrated in Fig. 4.4(d).

4.4 Current Considerations

The coil assembly generating the FFL field is supposed to stay static in space. The rotation as well as the translation of the FFL field is caused only by the variation of the applied currents. Hence, a detailed description of the required currents is needed. The choice of currents depends on the desired gradient strength achieved in the FFL field. The choice of current does also depend on the sensitivity of the used coils, as seen in Eq. (4.11) and Eq. (4.13). The sensitivity depends on the geometry of the setup and is computed by numerically solving the Biot-Savart integral [Jac99].

Since the fields $H^{a,\theta}$ and $H^{b,\theta}$ are each generated by a couple of Maxwell coil pairs, the currents and the sensitivities of these coils as well as the current in and the sensitivity of the Maxwell coil pair generating H^c are responsible for the gradient strength achieved in the FFL field. To generate the field provided by Eq. (7.7), with gradient strength g, the currents have to be chosen according to the former derived relations. Since an FFL field in z-direction shall be provided by Maxwell coil pairs in x- and y-direction, the currents need to be chosen according to Eq. (4.13). For $H^{a,\theta}$, this leads to

$$I_x^a = \frac{g}{3p^a} \cos 2\theta = -I_y^a, \tag{4.20}$$

where correspondingly

$$I_x^b = -\frac{g}{3p^b} \sin 2\theta = -I_y^b \tag{4.21}$$

is received for $H^{b,\theta}$. The currents in the coil pairs generating $H^{a,\theta}$ and $H^{b,\theta}$ are denoted I^a and I^b, while p^a and p^b are the corresponding coil sensitivities.

The dynamic part of the FFL selection field is hence generated by applying sinusoidal currents to the field generating coils taking the appropriate phase relations into account.

H^c is an FFP field in z-direction generated by a Maxwell coil pair in z-direction with coil sensitivity p^c leading to a current of

$$I_z^c = \frac{g}{p^c}. \tag{4.22}$$

Applying these currents to the field generating coils, specified in Fig. 4.4(a), Fig. 4.4(b), and Fig. 4.4(c), will lead to the desired rotating FFL field.

4.5 FFL Field Translation

The translation of the FFL is realized in a similar way as already discussed for FFP imaging. A spatially homogeneous 3D oscillating magnetic field is generated via three Helmholtz coil pairs with axes orientated along the x-, y-, and z-direction. In an ideal case, these fields are given by

$$H_{D,x}(t) = I_x(t)p_x\hat{e}_x$$

$$H_{D,y}(t) = I_y(t)p_y\hat{e}_y$$

$$H_{D,z}(t) = I_z(t)p_z\hat{e}_z, \tag{4.23}$$

where $\hat{e}_x$, $\hat{e}_y$, and $\hat{e}_z$ are the unit vectors of the Cartesian coordinate system. With a combination of these fields an arbitrary 3D trajectory can be applied. For the 3D case of FFL imaging in MPI, it is very likely that, similar to CT, the measurement will be performed in slices, while these slices are accordingly moved in the third dimension. Hence, only a 2D FFL trajectory will be considered. As introduced in section 3.2, the 2D FFL trajectory is of radial shape featuring a slow rotation provided by the selection field, and a fast rotation generated by the drive field.

The extension to 3D would then be realized via one additional Helmholtz coil pair moving the plain of interest in the remaining direction.

CHAPTER 5

A Field Free Line Field Demonstrator

5.1 Introduction

To transfer the theoretical results of the previous chapters to an experimental demonstration and validation of the dynamic FFL imaging concept, an FFL field demonstrator capable of generating an arbitrarily rotated and translated FFL field is implemented [EKS+11a]. This field demonstrator constitutes a proof of principle for magnetic field generation for FFL imaging in MPI. Performed measurements and achieved results motivate optimism towards experimental feasibility of dynamic FFL imaging in MPI.

The implemented FFL field demonstrator is capable of generating a complete radial FFL trajectory as needed for dynamic FFL imaging and as introduced in section 3.2. The field demonstrator is implemented, tested and evaluated with respect to magnetic field quality by comparing the achieved results to simulated data obtained from numerical evaluation of the Biot-Savart integral. It generates an FFL selection field with a gradient strength of $0.25\,\mathrm{Tm^{-1}}\mu_0^{-1}$. Hence, the field demonstrator is not meant for imaging but for proving experimental feasibility of magnetic field generation required for the implementation of a dynamic FFL imager in MPI. The coil arrangement chosen for implementation is analogue to the design introduced in Fig. 4.1.

Figure 5.1 Photo of the implemented FFL field demonstrator.

5.2 Materials and Methods

The implemented FFL field demonstrator is shown in Fig. 5.1. In this section, the construction process, the electromagnetic coils, their manufacturing and properties, as well as the process of measuring the magnetic fields of a complete dynamic FFL trajectory at distinct positions, are described.

5.2.1 FFL Field Demonstrator - Construction

The implemented setup consists of five Maxwell coil pairs according to the setup shown in Fig. 4.1. Two large coil pairs constitute the outer cube, while two smaller coil pairs, rotated by 45 degrees around the z-axis, form the inner cube. In addition, a Maxwell coil pair in z-direction, responsible for the static part of the field, is implemented.

The coils are mounted on a PVC holding. The holding was planned, designed and constructed using the computer-aided design (CAD) program SolidWorks (Dassault Systèmes). A construction drawing of the complete setup is illustrated in Fig. 5.2.

Figure 5.2 Illustration of the FFL field demonstrator, which was planned and constructed using the CAD program SolidWorks (Dassault Systèmes).

The coils need to be mounted on the holding. For that reason, the outer and the z-coils are winded with a PVC plate in the center as visible in Fig. 5.3. The PVC plate is screwed to the holding. The manufactured inner coils are glued on a PVC plate and are also screwed to the holding. This technique was chosen to ensure that the coils can be readjusted during the calibration process.

The field demonstrator is mounted on a wooden box and was constructed to be height adjustable as illustrated in Fig. 5.4. This way exact adjusting of the setup is ensured. Furthermore, inaccuracies due to changes of the environment can be readjusted.

In Fig. 5.5, the distances between the coil centers are specified for all three components, i.e. outer, inner, and z-coils.

Figure 5.3 The outer as well as the z-coils are winded with a PVC plate in center. The PVC plate is then screwed to the PVC mounting of the demonstrator.

Figure 5.4 The FFL demonstrator is mounted on a wooden box. The PVC mounting is constructed to be height adjustable.

5.2.2 FFL Field Demonstrator - Electromagnetic Coils

As already motivated, the implemented setup consists of five Maxwell coil pairs. The outer and inner coil pairs generate the dynamic part of the drive field and the z-coil pair is responsible for the static part of the selection field. To realize translation of the FFL field, the outer coil pairs are used as combined selection and drive field coils. The assembly generates a gradient strength of $0.25 \ \text{Tm}^{-1}\mu_0^{-1}$ perpendicular to the FFL.

The outer coils have an external diameter of 92 mm, an internal diameter of 50 mm, and the length, which is defined as the dimensions of the coil in direction of the coil axis in this work, amounts to 12 mm. The coils consist of 54 windings and have an inductance of 219 μH. The selection field current corresponds to Eq. (4.20) and is determined to be

$$I_{\text{outer}}^{\text{AC}} = 6.76 \ \text{A} \ \cos(2\theta_{\text{FFL}}) . \tag{5.1}$$

Their peak power loss is 4.79 W. For the FFL translation, an additional current, which is equally orientated in opposing coils, is applied to the outer coils. Doing so, a complete radial 2D trajectory is realized.

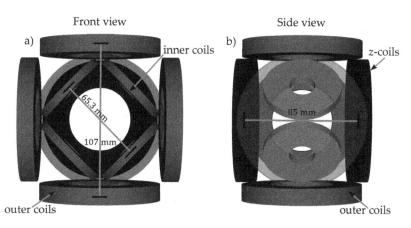

Figure 5.5 Front view (setup a)) and side view (setup b)) true to scale illustration of the implemented FFL field demonstrator with specifications on the distances of the coil centers.

The inner coils have an external diameter of 58 mm, an internal diameter of 19 mm as well as a length of 6 mm. They consist of 34 windings and their inductance is 38 μH. According to Eq. (4.21) the applied current is

$$I_{\text{inner}}^{\text{AC}} = 4.54 \text{ A sin}(2\theta_{\text{FFL}}) . \tag{5.2}$$

Each of these coils has a peak power loss of 0.74 W.

The z-coils have an external diameter of 92 mm, an internal diameter of 50 mm, a length of 16 mm, and consist of 72 windings. Their inductance amounts to 365 μH. They are responsible for the static part of the current

$$I_Z^{\text{DC}} = 9.1 \text{ A} . \tag{5.3}$$

This results in a power loss of 11.58 W.

Hence, the peak power loss of the implemented setup adds up to 42.32 W. The system is air-cooled to dissipate the generated heat.

The electromagnetic coils generating the fields needed for FFL imaging in MPI are manufactured especially for this purpose. They are made of litz wire, which is used to reduce the influence of the skin effect at high frequencies [Kad59]. High frequencies appear in the MPI drive field and have no importance regarding the FFL field demonstrator, since static measurements will be performed.

Table 5.1 Currents according to Eq. (5.1)-(5.3) and electrical power consumption of the field generating coils. The currents are determined to generate a gradient strength of $0.25 \, \mathrm{Tm^{-1}\mu_0^{-1}}$.

Coil	outer coil	inner coil	z-coil
External diameter / mm	92	58	92
Internal diameter / mm	50	19	50
Length /mm	12	6	16
Inductance /μH	219	38	365
Current (rotation) /A	$6.76 \cos(2\theta_{FFL})$	$4.54 \sin(2\theta_{FFL})$	9.2
Electrical power loss /Watt	$4.79 \cos^2(2\theta_{FFL})$	$0.74 \sin^2(2\theta_{FFL})$	11.58

However, the field demonstrator constitutes a test assembly verifying the simulated results on FFL imager designs presented in [KEB+10]. In an imaging experiment, litz wire will be used. It is therefore reasonable to use litz wire for the field generating electromagnetic coils of the field demonstrator.

The litz wire is winded into a special coil form, glued and compressed to achieve a high filling factor, i.e. ratio of copper to air, lacquer and silk.

The properties of the three coil types are summarized in Tab. 5.1

5.2.3 FFL Field Demonstrator - Measurement Process

Magnetic fields characterized by an FFL along six distinct angles have been generated, measured, and evaluated regarding the field quality by comparing it to simulated data. For each angle the appropriate currents in the outer and inner coils were determined using Eq. (5.1) and Eq. (5.2), while the currents in the z-coils were chosen according to Eq. (5.3). The currents were applied via Sorensen DLM 8-75 DC power supplies. A Gauss meter (LakeShore Model 475) with an axial Hall sensor was used to measure the z-component of the magnetic field, while a transversal sensor was used to obtain the x- as well as the y-component. To measure both field components, the transversal Hall probe was rotated by 90 degrees. As illustrated in Fig. 5.6 a), a quadratically shaped Hall probe mounting device ensured exact 90 degrees rotation. The device including the mountings for the axial as well as the transversal Hall probe was constructed in such a way, that the Hall sensor is located on precisely the same position for

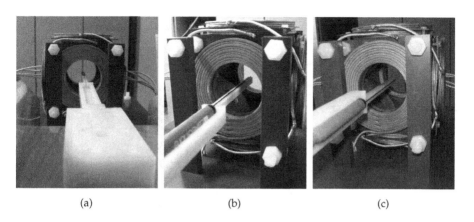

(a)   (b)   (c)

Figure 5.6 Mounting of the Hall sensors. Picture (a) shows the holding in the shape of a block with the quadratic cutout allowing for exact 90 degrees rotation of the probe. Picture (b) and (c) show the measurement of the x- and y-component, respectively.

both probes and after rotation of the transversal sensor. It was hence possible to measure all three components at one specific location by simply changing the retaining brackets. The 90 degree rotation of the transversal probe is illustrated in Fig. 5.6(b) (x-component) and Fig. 5.6(c) (y-component). The Hall sensor was moved by a robot (Iselautomation GmbH & Co. KG) through the FOV, which consists of 15×15 pixels and 28×28 mm^2.

5.3 Results

With the FFL field demonstrator it is possible to generate an arbitrarily rotated and translated FFL field. These fields were generated, measured, and evaluated with respect to magnetic field quality. The achieved results are presented in this section.

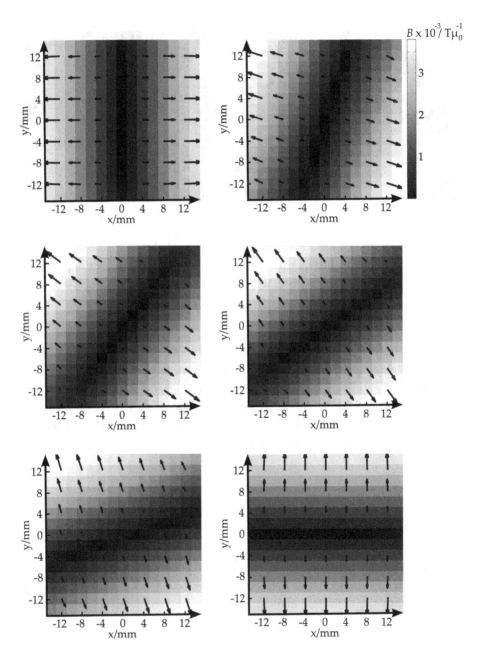

Figure 5.7 Absolute values of the measured magnetic fields of the FFL rotation, i.e. the FFL selection field.

5.3.1 FFL Rotation

Magnetic fields characterized by an FFL along six equidistant angles between 0 and 90 degrees are generated by the presented FFL field demonstrator are measured using a Hall sensor. The absolute values of the measured magnetic fields, in a region of $28 \times 28 \, \text{mm}^2$, are illustrated in Fig. 5.7. The occurrence and the rotation of the magnetic FFL is clearly visible. To compare these fields to simulated data, obtained via numerical evaluation of the Biot-Savart integral, the normalized root mean square deviation (NRMSD), the square root of the mean squared error divided by the range of the measured values, was computed. As listed in Tab. 5.2 it ranges from 1.39 % to 3.46 %. For the FFL orientated along 0 and 90 degrees the currents in the inner coils are zero. Since in this case only 3 of the 5 Maxwell coil pairs may cause errors, the NRMSD is likely to be small compared to the remaining four angles.

Considering the number of components contributing to the implemented setup the agreement of theory and experiment is very promising. Certainly, this experimental validation of the FFL concept underlines the feasibility of a dynamic FFL imaging device for the use in MPI.

5.3.2 FFL Translation

In addition to the rotation of the FFL a translation of the field has been performed, measured and evaluated with respect to simulated data. A displacement of 7.5 mm for the FFL orientated parallel to the x- and y-axis as well as a displacement of 4 mm for the FFL rotated by 45 degrees with respect to the x-axis has been realized. The measured results are illustrated in Fig. 5.8. Again, the NRMSD has been calculated with respect to simulated data. It amounts to 1.76 % for the horizontal translation, to 1.60 % for the diagonal translation and to 1.62 % for the vertical translation.

With these results, it is experimentally validated for the first time that a com-

Table 5.2 Normalized root mean square deviation of the measured FFL fields.

Angle in rad	0	$1/10\,\pi$	$1/5\,\pi$	$3/10\,\pi$	$2/5\,\pi$	$1/2\,\pi$
NRMSD in %	1.79	1.62	2.22	2.91	3.46	1.39

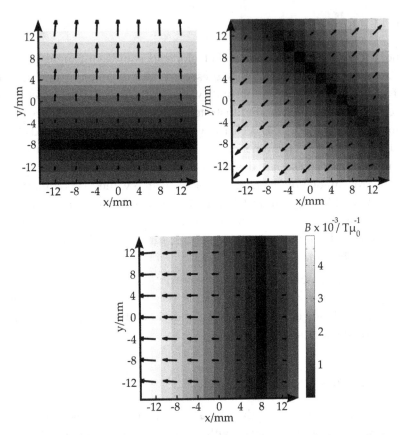

Figure 5.8 Absolute values of the measured fields of the FFL translation, i.e. combined selection and drive field generation.

plete dynamic FFL trajectory can be generated with the expected, low electrical power consumption.

5.3.3 Magnetic Field Quality

Approaching the scanner setup, the field strength along the FFL rises, i.e. the field homogeneity decreases. Of special importance is the field strength at which the SPIO particles used for imaging go into saturation and do therefore no longer contribute to the measured signal. This saturation field strength can

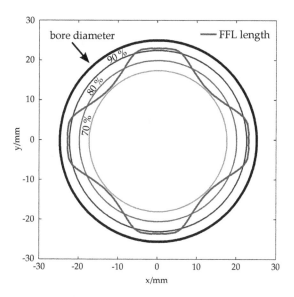

Figure 5.9 Length of the FFL at a gradient strength of $2.5 \text{ Tm}^{-1}\mu_0^{-1}$.

be defined as the field strength at which the particle magnetization reaches 90 % of its maximum value. Considering particles with a core diameter of 30 nm, the saturation field strength is $6.35 \text{ mT}\mu_0^{-1}$, according to the Langevin model (section 2.2.5).

To evaluate the field homogeneity along the FFL, the FFL length is introduced, which is identical to the distance to the center, at which the field strength along the FFL rises above the saturation field strength. This distance can be computed for every angle of the FFL. To examine the FFL homogeneity for a realistic MPI scanner using the implemented coil topology, the length of the FFL is computed for an FFL field with a gradient of $2.5 \text{ Tm}^{-1}\mu_0^{-1}$, which is commonly used in MPI experiments [WBG07]. The results are illustrated in Fig. 5.9. The FFL length is plotted in dependency of the FFL angle, where the FOV was chosen to be identical to the bore diameter of the z-coils, i.e. 50 mm. The grey circles show the spatial position of 90 %, 80 %, and 70 % of the FOV. The area that is enclosed by the FFL length represents the region where the field strength along the FFL does not exceed the saturation field strength. The length of the FFL and in turn the FFL homogeneity reaches its maximum of 92.4 % of the FOV at an angle of $(n \cdot \frac{\pi}{2} \pm 0.04)$ rad, for $n = 0, 1, 2, 3$, while the minimal length of the FFL amounts to

72.0 % of the FOV at angles $(2n + 1) \cdot \frac{\pi}{4}$ rad. Hence, the FFL homogeneity reaches its maximum at an angle close to the axes of the outer coils and its minimum at the axes of the inner coils. This is due to the larger size and distance of the outer coil pairs resulting in a more linear field profile.

The shape of the area in which the field strength is lower than the saturation field strength is roughly quadratic with rounded vertices. If a quadratic FOV is chosen for imaging, it would be beneficial to rotate the setup by 45 degrees so that the area of the highest field homogeneity optimally fits the shape of the FOV.

5.4 Discussion

The implementation of the presented FFL field demonstrator provides a proof of principle for magnetic field generation for dynamic FFL imaging in MPI. The field demonstrator represents an initial step towards feasibility of FFL imaging in MPI and was implemented for the exact same reason. Accounting for the simplicity of the technical implementation, the generated fields do agree to high extend with simulated data. The NRMSD for the rotated FFL fields ranges from 1.39 % to 3.46 %, where it almost continuously rises from the first to the fifth measured angle (compare Tab. 5.2). The NRMSD for the sixth angle is again much lower. The reason for this could be an interjection of the measurement between the fifth and the sixth measured angle. The system had the chance to cool down during this interjection, which had a positive impact on the achieved results. As a conclusion, system heating might be a significant error source and even better results could have been achieved, if an appropriate cooling concept would have been developed.

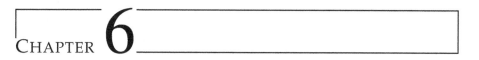

CHAPTER **6**

Scanner Efficiency and Magnetic Field Quality Analysis for Different Coil Topologies

Regarding the quality of a coil design for a dynamic FFL imager, two major aspects need to be considered. The first is the electrical power consumption, which mainly determines the requirements on the system cooling concept, which strongly influences the implementational effort. Hence, a coil design is needed, which provides a preferably low electrical power consumption.

The second important aspect searching for an optimal dynamic FFL imager design, is the magnetic field quality. The significance of magnetic field quality considerably increased, since it was shown in [KES+11] that efficient Radon-based reconstruction algorithms can be used for the FFL encoding scheme, which speeds up the reconstruction process. These reconstruction techniques will be introduced in chapter 7 and the influence of the magnetic field quality achieved with different designs on the quality of Radon-based reconstructed images will be analyzed.

In this chapter an FFL imager design is presented, which outperforms all formerly presented designs with respect to magnetic field quality as well as electrical power consumption. The presented design was therefore chosen for the implementation of the worlds first dynamic FFL imager, which is implemented as part of this thesis and is introduced in chapter 8. The presented simulation study therefore enabled a proof-of-principle for the dynamic FFL imaging concept.

Note, that the goal of the presented considerations on the FFL imager design is the implementation of a dynamic FFL mouse scanner. Of course, more complicated design options could be tested using the presented methods. Furthermore, an optimization of the current distribution with regard to an ideal FFL field could be applied. These steps, however, would lead to far more complicated coils topologies, which would not allow for implementation of such a design.

6.1 Dynamic FFL Selection Field Generation

This section provides an analysis of the coil topologies used for magnetic field generation of the FFL selection field. It was shown in [KSBB10] that 3 Maxwell coil pairs are needed to generate the selection field of an FFL imaging device. However, the field quality proofed to be optimal for a combination of 4 Maxwell coil pairs located at equidistant angles [KES+10a]. So far, circular coils have been considered, only. In this work, however, four different coil topologies, i.e. circular coils, curved circular coils, rectangular coils, and curved rectangular coils are tested regarding their ability for magnetic selection field generation of an FFL imaging device in MPI. The presented simulation study leads to a design, which provides - compared to circular coils - an improvement in magnetic field quality by a factor of almost five while decreasing electrical power consumption by a factor of almost four at the same time.

The analyzed scanner designs including their parameters are introduced in section 6.1.1. The analysis of the quality of the FFL scanning devices is organized in three parts. First, a current optimization is introduced in section 6.1.2, which provides the optimal choice of current with respect to magnetic field quality for each design. Using this method guarantees comparability of the presented results. In a second step, magnetic field quality of the selection field, i.e. the rotated FFL field, will be analyzed using different metrics, which are introduced especially for this purpose in section 6.1.3. Finally, the electrical power consumption of each designs is simulated in section 6.1.4.

The results will subsequently be summarized identifying the optimal coil design for dynamic FFL selection field generation in section 6.1.5. Since the major goal of the presented analysis is the implementation of the very first dynamic FFL mouse scanner, the results need to be validated with respect to implementation. For this purpose, a test assembly is implemented and the generated fields are measured and evaluated with respect to the simulations. The promising results are presented in section 6.1.6. A discussion will close this section.

6.1.1 Coil Topologies for Dynamic FFL Selection Field Generation

All dynamic FFL scanning devices introduced prior to [ESKB12] and this thesis consisted of circular coils, only [WGR+09, KSBB10, KES+10a]. To enhance the efficiency as well as magnetic field quality of dynamic FFL imagers, scanning devices consisting of four different coil topologies are simulated and analyzed. An illustration of the analyzed scanner designs of circular coils, curved circular coils, rectangular coils, and curved rectangular coils is presented in Fig. 6.1. The presented analysis starts with circular coils, since they can be considered as a measure for the improvement due to the standard use of these coils so far. The arrangement of the coils emanates from the design presented in [KES+10a] and experimentally validated with respect to magnetic field generation in [EKS+11a] (see chapter 5).

All presented simulations shall lead to the implementation of the first dynamic FFL mouse scanner presented in chapter 8. Therefore, the simulation parameters are chosen with regard to this scanner. The bore diameter of all scanning devices amounts to 40 mm to fit a receive coil assembly inside of the bore and spare room for a mouse. The gradient strength of all designs amounts to $1.5\,\mathrm{Tm}^{-1}\mu_0^{-1}$, which is also realized in the implemented scanner.

Note, that the imaging system presented in this thesis is air-cooled to reduce implementational effort and rather speed up the construction process to provide a proof of feasibility. This was not provided until the presented work. Making use of oil or water cooling would allow for a higher gradient strength. This can be realized in future dynamic FFL imaging devices.

The presented optimization methods refer to an FOV of 28×28 mm^2. Considering the fact that a receive coil assembly will be placed inside of the selection and drive field coils, this FOV covers the whole region interesting for an imaging experiment. The coil length was chosen to 9.2 mm and their thickness to 4.6 mm.

6.1.2 Current Optimization

To compare the magnetic field quality of a specific FFL imaging design, the deviation of the selection field from an ideal FFL field needs to be analyzed. In the presented dynamic FFL imaging method, rotation as well as translation of the FFL field is realized via varying the currents in the field generating coils only. No mechanical movement is required at any time. However, the remaining question is, how to find an appropriate process for optimizing the coil currents, which is comparable for all designs and provides the optimal field quality achievable with a specific design. The current optimization methods are presented in this section.

As introduced in Eq. (2.29), each coil has a particular sensitivity p. For a scanner design of $i = 1, ..., N$ different electromagnetic coils, the resulting magnetic field is then composed of a superposition of the coil sensitivities weighted with the applied current

$$H(r) = \sum_{i=1}^{N} H_i(r) = \sum_{i=1}^{N} I_i p_i(r).$$ (6.1)

The sensitivity of each coil in a scanning design can be predetermined using a numerical evaluation of the Biot-Savart integral [Jac99]. Furthermore, an arbitrary target field H_{target} can be chosen. In the presented case, this is an ideal rotating FFL field with a gradient strength of $1.5\,\text{Tm}^{-1}\mu_0^{-1}$.

Knowing the sensitivity of each coil as well as the desired target field a minimization problem can be solved

$$\arg\min_{I} \|Ip - H_{\text{target}}\|_2^2$$ (6.2)

Applying this method leads to the optimal current for each coil of the scanner design.

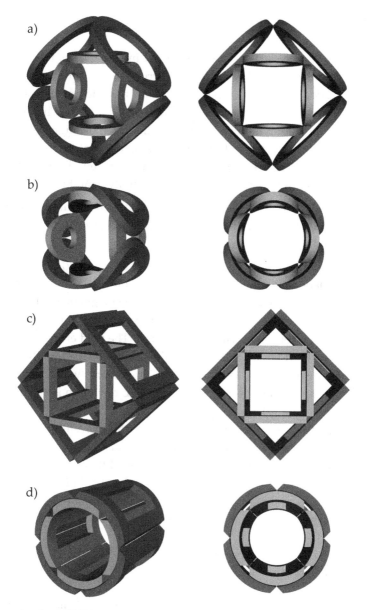

Figure 6.1 Analyzed FFL scanner designs consisting of four different coil topologies, i.e. circular coils in a), which constitute the standard, curved circular coils b), rectangular coils c), and curved rectangular coils d).

6.1.3 Magnetic Field Quality

Magnetic field quality analysis gained special importance as it was shown in [KES⁺11] that powerful Radon-based reconstruction algorithms can be applied using a line encoding scheme. These algorithms assume line detection and do not account for field inhomogeneities in their simplest form. Therefore, a high field homogeneity along the FFL as well as parallel to its alignment is of great importance for the image quality achieved using efficient Radon-based reconstruction algorithms. A measure for the FFL homogeneity is presented, which provides the possibility to predict the imaging ability of an FFL scanner design. The magnetic field quality achieved for the presented dynamic FFL scanner designs is analyzed. However, the presented evaluation methods can be applied to an arbitrary dynamic FFL imager.

To compare various FFL scanner geometries, a metric is needed to evaluate the achieved magnetic field quality. To do so, the normalized root mean square deviation (NRMSD) can be considered. However, the magnetic field quality close to the FFL, where SPIO tracer particles contribute to the acquired signal, is of special importance for the FFL imaging method. The field errors in a region far away from the FFL do not significantly influence the image quality. Therefore, the mean FFL deviation $\bar{\epsilon}_{FFL}$ is introduced in this work, which names the average field strength along the FFL. The mean FFL deviation is calculated for a whole 360 degree FFL rotation in 1 degree steps. In this way, a map of the magnetic field quality depending on the position of the FFL within the scanner is provided. The influence of the shape of the electromagnetic coils used for magnetic field generation can be evaluated by means of this metric. By averaging the mean FFL deviation with respect to the angle of the FFL, a single number, the mean FFL deviation number $\bar{\sigma}_{FFL}$ is provided to compare the magnetic field quality of different FFL scanner assemblies.

6.1.3.1 FFL Deviation

A measure is needed to evaluate and compare the magnetic field quality of FFL scanning devices. As already discussed, the NRMSD is not an optimal choice for the presented application. In MPI, only particles in the close vicinity of a field free region, i.e. the FFP or as in the presented case the FFL, contribute to the measured signal. Hence, we are mainly interested in magnetic field quality in this specific region.

Therefore, a measure for magnetic field quality evaluation in dynamic FFL imaging in MPI is introduced, the FFL deviation $\epsilon_{FFL}(r, \phi)$. The FFL deviation $\epsilon_{FFL}(r, \phi)$ is the absolute value of the magnetic field at a specific spatial position along the FFL, where r denotes the distance of this spatial position along the FFL from the center of the scanner setup and ϕ is the angle of the FFL with respect to the x-axis. In an ideal case, of course, the field strength along the FFL should be zero in all three components. This is, however, not the case for a realistic scanner device as presented in this work. Furthermore, the FFL deviation is calculated for a region of interest (ROI) covering the whole bore diameter of the scanner. Considering only a small region in the center of the scanner setup does not suffice to make a statement about the magnetic field quality of a specific design, since deviation grows approaching the scanner setup. It is desirable to use a region as large as possible for imaging, since power consumption rises for larger imaging devices.

The FFL deviation is defined as

$$\epsilon_{FFL}(r, \phi) = |H(r, \phi)| - \underbrace{|H_{ideal}(r, \phi)|}_{0 \text{ along the FFL}} = |H(r, \phi)|. \tag{6.3}$$

The field error is hence the field strength itself regarding positions on the FFL. The FFL deviation $\epsilon_{FFL}(r, \phi)$ is simulated using a numerical evaluation of the Biot-Savart integral for a whole 360 degree FFL rotation at a discrete number of spatial positions n_r and angular positions n_ϕ. The results are plotted in figure 6.2 for circular coils, in figure 6.3 for curved circular coils, in figure 6.4 for rectangular coils, and in figure 6.5 for curved rectangular coils.

The dimension of the errors does already give an impression of the magnetic field quality of the various designs. The errors are smallest for curved rectangular coils and highest for curved circular coils. Their shape represents a map of the field inhomogeneities introduced by the four scanner designs. The error consistently rises along the axes of the inner, smaller coils.

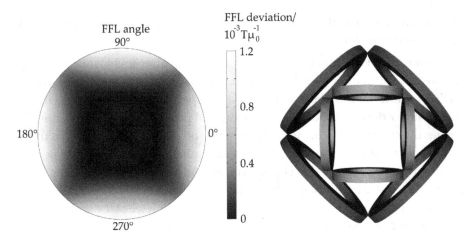

Figure 6.2 FFL deviation for an assembly of circular coils generating the FFL selection field.

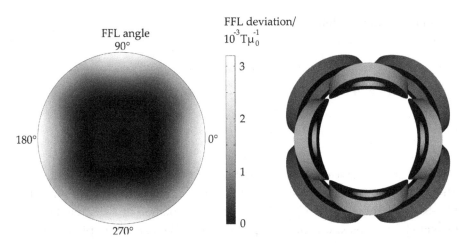

Figure 6.3 FFL deviation for an assembly of curved circular coils generating the FFL selection field.

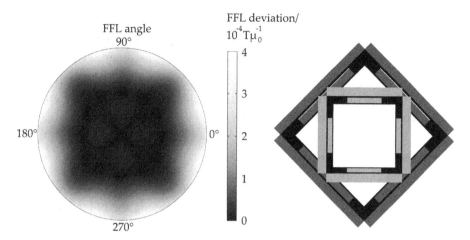

Figure 6.4 FFL deviation for an assembly of rectangular coils generating the FFL selection field.

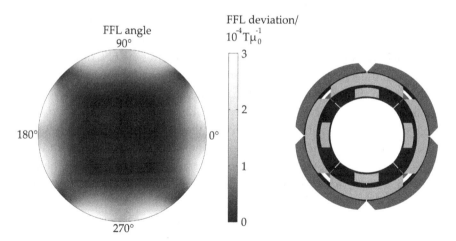

Figure 6.5 FFL deviation for an assembly of curved rectangular coils generating the FFL selection field.

This can also be explained from a phenomenological point of view, since enhancing the diameter of a coil compared to the regarded FOV will improve the field quality. This effect will also be discussed in the subsequent section.

However, a more generic representation of the errors is needed to actually compare the various designs and analyze the influence of the coil shape and arrangement on the magnetic field quality.

6.1.3.2 Mean FFL Deviation

For a more generic error analysis, the mean FFL deviation $\bar{e}_{FFL}(\phi)$ is introduced, which is the average of the magnetic field strength along the FFL regarding a given FOV for one specific FFL angle ϕ

$$\bar{e}_{FFL}(\phi) = \frac{1}{n_r} \sum_{l=1}^{n_r} |H(\phi, r_l)| . \tag{6.4}$$

The mean FFL deviation is calculated for a whole 360 degree FFL rotation in 1 degree steps. The mean FFL deviation in dependency on the FFL angle allows for making a statement about the influence of the coil arrangement on the magnetic field quality. As visible in Fig. 6.2-6.5, showing the FFL deviation, the orientation of the axes of the coils is of great importance for the magnetic field quality. Due to the fact that the outer coils are larger than the inner ones, the magnetic field quality is higher at the axes of the outer coils. For a given FOV, the field homogeneity or the homogeneity of the field gradient, respectively, is increased by increasing the size of the field generating coils. Ideal FFL and FFP fields, as derived in chapter 4, are only present in a very small area in the center of the scanner setup. With increasing size of the FOV, field inhomogeneities need to be accepted. The same holds for a given FOV and an increase in the size of the field generating coils, which increases the field homogeneity. The same effect also accounts for the increase in field quality obtained leaving the concept of a scanner consisting of a ring of coils.

The results are illustrated in Fig. 6.6 for circular coils, in Fig. 6.7 for curved circular coils, in Fig. 6.9 for rectangular coils, and in Fig. 6.8 for curved rectangular coils, which yield the smallest field errors.

To compare these results, Fig. 6.10 illustrates the mean FFL deviation in dependency of the FFL angle for a consistent scale. In this plot, it is clearly visible that curved rectangular coils are the optimal choice for the implementation of an FFL scanning device with respect to magnetic field quality.

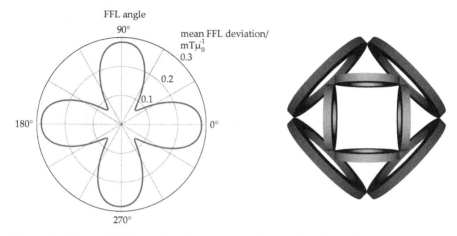

Figure 6.6 Mean FFL deviation for an assembly of circular coils generating the FFL selection field. The influence of the arrangement of the coils on the achieved magnetic field quality is clearly visible in the decrease of the mean FFL deviation at the axes of the outer selection field coils.

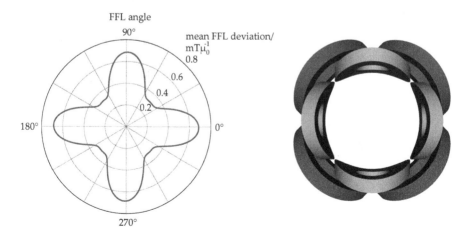

Figure 6.7 Mean FFL deviation for an assembly of curved circular coils generating the FFL selection field.

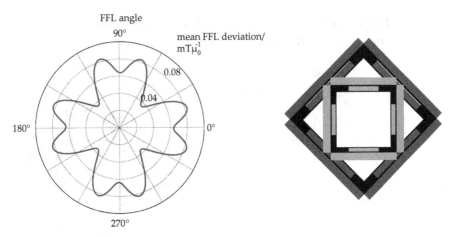

Figure 6.8 Mean FFL deviation for an assembly of rectangular coils generating the FFL selection field.

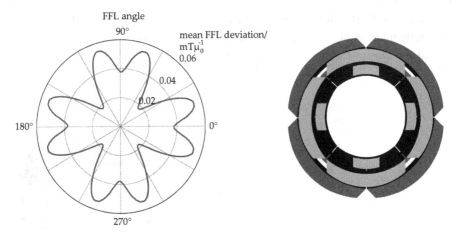

Figure 6.9 Mean FFL deviation for an assembly of curved rectangular coils generating the FFL selection field.

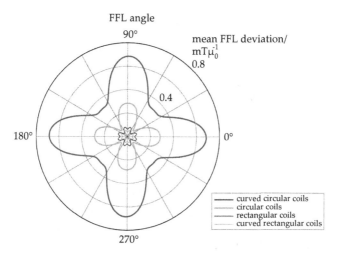

Figure 6.10 Mean FFL deviation of all four considered coil topologies in a consistent scale.

6.1.3.3 Mean FFL Deviation Number

To evaluate the enhancement in magnetic field quality achieved by using curved rectangular coils instead of the standard choice of circular coils the mean FFL deviation number is defined

$$\bar{\sigma}_{FFL} = \frac{1}{n_\phi} \sum_{l=1}^{n_\phi} \bar{\epsilon}_{FFL}(\phi_l), \tag{6.5}$$

which allows for direct comparison of the magnetic field quality of a specific FFL scanning devices by using one single number. The mean FFL deviation number results from averaging the mean FFL deviation with respect to the angular sampling number, n_ϕ. The results normalized with respect to the standard results achieved for circular coils

$$\bar{\sigma}_{FFL,\,norm} = \frac{\bar{\sigma}_{FFL}}{\bar{\sigma}_{FFL,\,circular\,coils}} \tag{6.6}$$

are specified in Tab. 6.1.

Table 6.1 Normalized mean FFL deviation number.

Coil topology	$\bar{\sigma}_{FFL,\,norm}$
Circular	1
Curved Circular	2.45
Rectangular	0.33
Curved Rectangular	0.21

6.1.4 Electrical Power Consumption

The analysis regarding the magnetic field quality clearly reveals the FFL scanner design consisting of curved rectangular coils as the optimal choice. Despite of the importance of magnetic field quality, another crucial aspects needs to be taken into account, the electrical power consumption.

Since the goal of the presented analysis is to find the best possible design for the implementation of a dynamic FFL imager, the electrical power consumption, which determines the required cooling concept, is at least as important as the magnetic field quality. The best possible compromise between these two aspects needs to be sought.

The electrical power consumption of the scanner designs can be simulated regarding the current optimization introduced in the section 6.1.2 as well as the properties of the litz wire, which will be used for implementation. Litz wire consists of many very thin twisted conductor strands. With the pressing process used during implementation, a filling factor of $\rho_f = 0.53$ of pure copper inside of the coils can be achieved. Litz wire is used to prevent influence of the skin effect as well as due to the fact that it alleviates the manufacturing process. Furthermore, the resistivity ρ of copper is needed. It amounts to $1.72 \times 10^{-8}\,\Omega m$ at a temperature of 20 degree and is a temperature dependent material constant. The resistance R, which is needed to simulate the electrical power consumption is related to the resistivity via

$$R = \rho \frac{l}{A}, \tag{6.7}$$

where A is the cross section of the conductor and l its length, both parameters are known for each coil of the simulated scanner designs. In the presented case, where we assume the coils to be manufacture of litz wire, the filling factor needs

Table 6.2 Simulated normalized electrical power consumption of the presented FFL scanner designs.

Coil topology	normalized electrical consumption
Circular	1
Curved Circular	0.23
Rectangular	1.34
Curved Rectangular	0.28

to be taken into account. The effective cross section of the conductor therefore amounts to

$$A_{\text{effective}} = \rho_f A . \tag{6.8}$$

The electrical power consumption can then be determined using

$$P = I^2 R = I^2 \rho \frac{l}{\rho_f A} . \tag{6.9}$$

The results for the total electrical power consumption are again given in absolute manner as well as normalized with respect to the standard result obtained for the use of circular coils (Tab. 6.2).

These results are very promising, since they identify curved rectangular coils not only as the coil design leading to the best magnetic field quality but also providing a considerably lower power consumption than circular coils. Only the curved circular coils have an even lower electrical power consumption, as discussed in the previous section, however, they do result in the worst magnetic field quality.

6.1.5 Summary

The FFL scanner design of curved rectangular coils constitute the optimal match of high magnetic field quality as well as low electrical power consumption. The presented results predict considerable improvement in magnetic field quality for the use of curved rectangular coils by reducing the mean FFL deviation number by a factor of 4.87 compared to circular coils. At the same time, the electrical power consumption is reduced by a factor of 3.57. These promising results guide the way towards implementation of the dynamic FFL scanner presented in chapter 8.

6.1.6 Experimental Validation

The presented analysis on coil designs in this chapter aims on preparing the implementation of the first dynamic FFL mouse scanner. Due to that, the presented simulation results need to be validated with respect to practical feasibility. As mentioned in the beginning of this chapter, only coil topologies, which are considered to be feasible with respect to implementation, are considered in the presented simulation study. However, it is obvious that the implementation of curved rectangular coils hold more challenges than the implementation of conventional circular coils. Hence, it needs to be validated, whether implementational errors occur for the use of curved rectangular coils, which deteriorate the achieved magnetic field quality. In that case, the simulated results would need to be reconsidered and extended by a parameter taking implementational errors into account.

A test assembly of four curved rectangular coils (Fig. 6.11(a)) was implemented and the resulting field was measured and analyzed. To do so, the exact same measurement instrumentation was used as for the evaluation of the fields generated with the FFL field demonstrator presented in chapter 5. Since the field demonstrator is constructed of circular coils, this data can be used as a reference.

The curved rectangular coils are manufactured using a customized peek pressing form illustrated in Fig. 6.11(b).

An FFL field in z-direction, i.e. through the scanner axis was generated and measured. The result is shown in Fig. 6.12. As for the fields generated with the field demonstrator the normalized root mean square deviation (NRMSD) with respect to the simulated field was calculated. It amounts to 2.67% and does hence not considerably differ from the NRMSD of the fields generated with circular coils. This is a very promising result with respect to the implementation of an FFL mouse scanner. Due to these results, the selection field coils of the first dynamic FFL scanner, which will be presented in chapter 8 of this thesis will be implemented in the shape of curved rectangular coils.

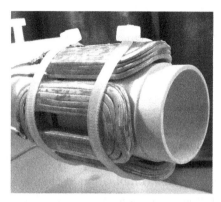

(a) Test assembly of curved rectangular coils implemented for validation of the simulated results.

(b) Customized peek form for the manufacturing curved rectangular coils.

Figure 6.11 Implemented test assembly.

6.1.7 Discussion

The presented analysis of the selection field generation for a dynamic FFL imaging system aims on the implementation of the first dynamic FFL mouse scanner. It is possible to determine the exact position of the current flow needed for an optimal field quality. This was done for the receiving coil of the scanner as will be presented in chapter 8. However, it is of great importance to consider not only the field quality, but also the electrical power consumption and on top of that feasibility with respect to implementation. Due to these reasons, the four

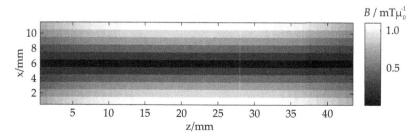

Figure 6.12 FFL field generated with the test assembly of curved rectangular coils.

presented coil topologies are analyzed and compared.

The presented methods can be applied to any arrangement of electromagnetic coils and an arbitrary target field.

6.2 FFL Drive Field Generation

Drive field generation for dynamic FFL imaging in MPI considerably differentiates from selection field generation. The drive field is a spatially homogeneous, time-varying magnetic field generated via a combination of Helmholtz coil pairs. Using two perpendicular Helmholtz coil pairs, an arbitrary 2D trajectory can be realized. Hence, the radial 2D FFL drive field is realized by such a simple combination of Helmholtz coils pairs. FFL drive field generation does not considerably differ from FFP drive field generation. The difference lies in the applied current, which is responsible for the shape of the trajectory. As a consequence, the knowledge gained from FFP drive field generation can directly be applied for FFL drive field generation.

To design the FFL drive field coils various aspects need to be taken into account. To ensure high magnetic field quality the drive field coils would need to be located as far away from the FOV as possible. At the same time, however – as for the selection field coils – the electrical power consumption needs to be considered. However, these are not the only aspects important for the implementation of the drive field coils. Mounting them in an area outside of the selection field coils, the AC current applied to them would couple in the selection field coils and would lead to field inhomogeneities. To reduce this coupling effect and to lower the power consumption, the drive field coils are located within the assembly of selection field coils. The exact implementation process is described in chapter 8.

CHAPTER 7

Efficient Reconstruction Algorithms

Reconstruction in MPI in most cases requires a time consuming calibration scan, from which the system function can be derived (see section 2.6). The system function is a matrix containing the signal caused by a delta probe, which is moved through the whole FOV. In a second step, the linear system of equations containing the relation between the particle distribution and the signal received during an MPI measurement is solved (see section 2.7). This method holds for FFP as well as for FFL imaging in MPI.

However, in the case of data acquisition using an FFL, more efficient reconstruction algorithms arise, since the data can be transformed into Radon space [KES+11]. Transforming the measured MPI signal into Radon space opens up the possibility to benefit from well known and powerful reconstruction algorithms like for instance the filtered backprojection (FBP).

FFL reconstruction considerations are closely linked to the optimization of the scanner geometry concerning magnetic field quality, since transforming the measured signal into Radon space demands high magnetic field homogeneity along the FFL as well as parallel to its alignment.

In this section, three major aspects will be discussed. First, it is proven that the signal measured by an FFL imaging device can indeed be transformed into Radon space and therefore can be reconstructed by using the FBP. Second, reconstructed images for ideal and realistic scanners are compared. And third, a simulation study on the influence of the magnetic field quality of an FFL imaging device on the image quality achieved when performing efficient Radon-

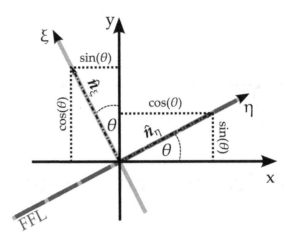

Figure 7.1 Coordinate notation for the FFL rotation

based reconstruction is performed and evaluated. It is shown that the optimization steps concerning the scanner design presented in chapter 6 do not only considerably improve power consumption, but do also reduce artifacts introduced when using Radon-based reconstruction algorithms.

7.1 Theory

In the subsequent section a theoretical derivation for the FFL signal transformation into Radon space is presented. The applied methods are based on [KES+11].

7.1.1 Magnetic Fields

As discussed in chapter 4, FFL imaging as introduced in this thesis requiresüüüüüüü a superposition of two different magnetic field types, the selection field and the drive field. In the case of FFL imaging, the selection field is an FFL field, which is rotated with frequency f_S. The drive field is a spatially homogeneous, time varying field oscillating with frequency f_D perpendicular to the FFL. In the following, these two fields are derived. It is convenient to introduce a rotated coordinate system first, which will simplify the derived expressions.

7.1.1.1 Coordinate Transformation

Since the FFL is rotated with frequency f_S, a rotated coordinate system is introduced (Fig. 7.1). The coordinate ξ is orientated perpendicular to the FFL and hence marks the direction of the FFL drive field, while η points in direction of the FFL alignment. The angle between η and the x-axis, or ξ and the y-axis respectively, is denoted θ. The coordinate transformations are given by

$$\xi = -x\sin\theta + y\cos\theta \tag{7.1}$$
$$x = -\xi\sin\theta + \eta\cos\theta \tag{7.2}$$
$$\eta = \ x\cos\theta + y\sin\theta \tag{7.3}$$
$$y = \ \xi\cos\theta + \eta\sin\theta. \tag{7.4}$$

The unit vectors in direction of the rotated coordinates are hence determined to be

$$\hat{n}_\xi = \begin{pmatrix} -\sin\theta \\ \cos\theta \end{pmatrix} \tag{7.5}$$

perpendicular to the FFL and

$$\hat{n}_\eta = \begin{pmatrix} \cos\theta \\ \sin\theta \end{pmatrix} \tag{7.6}$$

pointing in the direction of the FFL.

7.1.1.2 Selection Field

A theoretical derivation of the FFL selection field is provided in chapter 4. In Eq. (4.14) the FFL selection field is specified. It equals an FFL field rotated by an arbitrary angle θ with respect to the x-axis. Choosing the original FFL field to be $H^{\text{FFL},x} = \text{diag}\{0,-1,1\}$ for simplicity in the subsequent derivations, Eq. (4.14) can be transformed into

$$H_S^{\text{FFL},\theta}(r) = g \begin{pmatrix} \cos\theta & -\sin\theta & 0 \\ \sin\theta & \cos\theta & 0 \\ 0 & 0 & 1 \end{pmatrix} \begin{pmatrix} 0 & 0 & 0 \\ 0 & -1 & 0 \\ 0 & 0 & 1 \end{pmatrix} \begin{pmatrix} \cos\theta & \sin\theta & 0 \\ -\sin\theta & \cos\theta & 0 \\ 0 & 0 & 1 \end{pmatrix} r$$

$$= g \begin{pmatrix} -\sin^2\theta & \sin\theta\cos\theta & 0 \\ \sin\theta\cos\theta & -\cos^2\theta & 0 \\ 0 & 0 & 1 \end{pmatrix} \begin{pmatrix} x \\ y \\ z \end{pmatrix}$$

$$= g \begin{pmatrix} -x\sin^2\theta + y\sin\theta\cos\theta \\ x\sin\theta\cos\theta + y\cos^2\theta \\ z \end{pmatrix}$$

$$= g\,(x\sin\theta - y\cos\theta) \begin{pmatrix} -\sin\theta \\ \cos\theta \\ z \end{pmatrix}. \tag{7.7}$$

In the case of an FFL field located in the center of the scanner setup, i.e. in the xy-plane, this is

$$H_S^{\mathrm{FFL},\theta}(x,y) = g\,(x\sin\theta - y\cos\theta) \begin{pmatrix} -\sin\theta \\ \cos\theta \\ 0 \end{pmatrix}. \tag{7.8}$$

In terms of the rotated coordinates introduced in Eq. 7.2 and Eq. 7.4, the selection field can be expressed via

$$H_S^{\mathrm{FFL},\theta}(\xi) = -g\,\xi\,\hat{n}_\xi. \tag{7.9}$$

Hence, the direction of the FFL selection field is given by the unit vector oriented perpendicular to the FFL. Furthermore, this representation of the FFL selection field shows that the field is constant along any line parallel to the FFL for a given FFL angle θ, since ξ is constant along such a line and $H_S^{\mathrm{FFL},\theta}$ does not depend on η.

7.1.1.3 Drive Field

Reconstructing FFL images demands not only for FFL rotation via the selection field, but also for FFL translation via the drive field. The drive field is a spatially homogeneous, time varying magnetic field. It is orientated perpendicular to the FFL at any point in time and therefore needs to be rotated according to the FFL

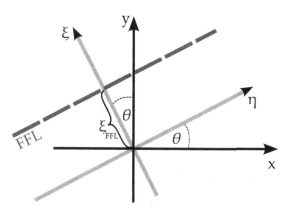

Figure 7.2 Coordinate notation for the FFL translation.

rotation. Its direction is therefore given by the unit vector $\hat{n}_\xi$. The FFL drive field can be described by

$$H_D^{\text{FFL},\theta}(\xi, t) = A_D \Lambda(t) \hat{n}_\xi. \tag{7.10}$$

The amplitude of the drive field is given by A and $\Lambda(t)$ denotes the excitation function, which is conventionally chosen to be sinusoidal in MPI, for both, FFP as well as FFL imaging. It is given by

$$\Lambda(t) = \cos(2\pi f_D t), \tag{7.11}$$

where f_D is the drive field frequency.

7.1.1.4 Total Magnetic Field

The superposition of the selection field $H_S^{\text{FFL},\theta}$ and the drive field $H_D^{\text{FFL},\theta}$ constitutes the total magnetic field to be considered

$$H^{\text{FFL},\theta}(\xi, t) = H_S^{\text{FFL},\theta}(\xi) + H_D^{\text{FFL},\theta}(\xi, t) = (A\Lambda(t) - g\xi)\hat{n}_\xi. \tag{7.12}$$

Along the FFL, the field is zero

$$0 = A\Lambda(t) - g\xi_{\text{FFL}} \tag{7.13}$$

and the FFL displacement ξ_{FFL} (Fig 7.2) is hence given by

$$\xi_{\text{FFL}} = -\frac{A\Lambda(t)}{g}. \tag{7.14}$$

The FFL is translated with a displacement of $\left[-\frac{A}{g}, \frac{A}{g}\right]$.

7.1.2 Particle Magnetization

The magnetization of an SPIO particle distribution has been introduced in Eq. (2.7) and depends on the SPIO particle concentration c and on the mean magnetic moment of the SPIO particle distribution $\bar{\mu}$

$$M(x, y, t) = c(x, y)\bar{\mu}(x, y, t). \tag{7.15}$$

Neglecting relaxation effects, the magnetic moments of the SPIO particles and hence the mean magnetic moment of a distribution of the same will point in direction of the total magnetic field given in Eq. (7.12). The mean magnetic moment $\bar{\mu}$ expressed in the rotated coordinates introduced in Eq. (7.2) and Eq. (7.4) is therefore given by

$$\bar{\mu}(\xi, \eta, t) = \bar{\mu}\left(H^{\mathrm{FFL},\theta}(\xi, t)\right) \operatorname{sgn}\left(\boldsymbol{H}^{\mathrm{FFL},\theta}(\xi, t)\right) \hat{n}_\xi, \tag{7.16}$$

where $\hat{n}_\xi$ denotes the unit vector in direction of $\boldsymbol{H}^{\mathrm{FFL},\theta}$ (compare Eq. (7.12)), and $\bar{\mu}$ is the modulus of the mean magnetic moment. Furthermore, the mean magnetic moment $\bar{\mu}$ depends on the modulus of the magnetic field, which is given by

$$H^{\mathrm{FFL},\theta}(\xi, t) = |A\Lambda(t) - g\xi| \tag{7.17}$$

according to Eq. (7.12). Hence, Eq. (7.17) transforms into

$$\bar{\mu}(\xi, t) = \bar{\mu}(|A\Lambda(t) - g\xi|) \operatorname{sgn}(A\Lambda(t) - g\xi) \hat{n}_\xi. \tag{7.18}$$

To further simplify this equation, a closer look at the symmetry characteristics of the mean magnetic moment needs to be taken. The symmetry corresponds to the characteristics of the Langevin function, which is given by $\mathcal{L}(\alpha) = \coth(\alpha) - \frac{1}{\alpha}$ and fulfills $f(-x) = -f(x)$.

Performing a case-by-case analysis of $\bar{\mu}(|A\Lambda(t) - g\xi|) \operatorname{sgn}(A\Lambda(t) - g\xi)$ yields

1. for $A\Lambda(t) - g\xi > 0$, $\operatorname{sgn}(A\Lambda(t) - g\xi) = 1$ and hence

$$\bar{\mu}(|A\Lambda(t) - g\xi|)\operatorname{sgn}(A\Lambda(t) - g\xi) = \bar{\mu}(A\Lambda(t) - g\xi). \tag{7.19}$$

2. for $A\Lambda(t) - g\xi < 0$ and hence $\operatorname{sgn}(A\Lambda(t) - g\xi) = -1$

$$\begin{aligned}
\bar{\mu}(|A\Lambda(t) - g\xi|)\operatorname{sgn}(A\Lambda(t) - g\xi) &= -\bar{\mu}(|A\Lambda(t) - g\xi|) \\
&= \bar{\mu}(-|A\Lambda(t) - g\xi|) \\
&= \bar{\mu}(A\Lambda(t) - g\xi).
\end{aligned} \tag{7.20}$$

The mean magnetic moment of a distribution of SPIO particles is hence given by

$$\bar{\mu}(\xi, t) = \bar{\mu}(A\Lambda(t) - g\xi)\hat{n}_\xi. \tag{7.21}$$

As it is the case for the magnetic field, the mean magnetic moment is constant along a line parallel to the FFL.

The magnetization response of a distribution of SPIO particles to the external magnetic field specified in (7.12) expressed in the rotated coordinates ξ and η is therefore given by

$$M(\xi, \eta, t) = c(-\xi \sin\theta + \eta \cos\theta, \xi \cos\theta + \eta \sin\theta)\bar{\mu}(A\Lambda(t) - g\xi)\hat{n}_\xi. \tag{7.22}$$

7.1.3 Induced Signal

According to Eq. (2.28), the voltage induced in a receive coil due to the change in magnetization of a distribution of SPIO particles is described by

$$u^\theta(t) = -\mu_0 \int_\Omega \frac{\partial}{\partial t} M(x, y, t) p(x, y) \, dx \, dy. \tag{7.23}$$

Assuming an ideal receive coil sensitivity p, which is homogeneous in space, and inserting Eq. (7.22), the induced voltage can be expressed as

$$u^\theta(t) = -\mu_0 \int_{\mathbb{R}^2} \frac{d}{dt} c(-\xi \sin\theta + \eta \cos\theta, \xi \cos\theta + \eta \sin\theta) \bar{\mu}(A\Lambda(t) - g\xi) \, \hat{n}_\xi p \, d\xi \, d\eta. \tag{7.24}$$

The inner product $\hat{n}_\xi p$ of the direction of the FFL movement and the receive coil sensitivity vanishes in the case where these two vectors are perpendicular. The scalar

$$q^\theta := \hat{n}_\xi p \tag{7.25}$$

is defined and Eq. (7.24) transforms into

$$u^\theta(t) = -\mu_0 q^\theta \int_{\mathbb{R}^2} \frac{d}{dt} c(-\xi \sin\theta + \eta \cos\theta, \xi \cos\theta + \eta \sin\theta) \bar{\mu}(A\Lambda(t) - g\xi) \, d\xi \, d\eta. \tag{7.26}$$

Taking a look at the definition of the Radon transform [Rad17]

$$\mathcal{R}(f)(\phi, r) := \int_{\mathbb{R}} f(r\cos\phi + t\sin\phi, r\sin\phi - t\cos\phi) dt \qquad (7.27)$$

and introducing a phase shift of $\pi/4$ for convenience due to the FFL angle obtaining $\cos(\phi + \frac{\pi}{4}) = -\sin(\phi)$ and $\sin(\phi + \frac{\pi}{4}) = \cos(\phi)$ leads to

$$\mathcal{R}(c)(\theta, \xi) := \int_{\mathbb{R}} c(-\xi\sin\theta + \eta\cos\theta, \xi\cos\theta + \eta\sin\theta) \, d\eta. \qquad (7.28)$$

The induced signal specified in Eq. (7.26) can thus be expressed by means of the Radon transform of the SPIO particle concentration c

$$u^\theta(t) = -\mu_0 q^\theta \int_{\mathbb{R}} \frac{d}{dt}\overline{\mu}(A\Lambda(t) - g\xi)\mathcal{R}(c)(\theta, \xi) \, d\xi. \qquad (7.29)$$

With

$$\frac{d}{dt}\overline{\mu}(A\Lambda(t) - g\xi) = \overline{\mu}'(A\Lambda(t) - g\xi)A\Lambda'(t) \qquad (7.30)$$

the induced voltage transforms into

$$u^\theta(t) = -\mu_0 q^\theta A\Lambda'(t) \int_{\mathbb{R}} \overline{\mu}'(A\Lambda(t) - g\xi)\mathcal{R}(c)(\theta, \xi) \, d\xi. \qquad (7.31)$$

With the definition of a convolution

$$(f_1 * f_2)(t) = \int_{\mathbb{R}} f_1(\tau) f_2(t - \tau) \, d\tau \qquad (7.32)$$

the induced signal can be expressed as

$$u^\theta(t) = q^\theta A\Lambda'(t) \ (\tilde{\mu} * \mathcal{R}(c)(\theta, \cdot))\left(\frac{A}{g}\Lambda(t)\right), \qquad (7.33)$$

where

$$\tilde{\mu}(x) := -\mu_0\overline{\mu}'(gx) \qquad (7.34)$$

is the convolution kernel.

7.1.4 Signal Transformations

To take advantage of Radon-based reconstruction mechanisms, further signal transformation is needed to compensate for various phenomena. These phenomena as well as the mathematical description of the same are described in the following.

7.1.4.1 Receive Channel Normalization

In a first step, the signal needs to be normalized with respect to the receive coil sensitivity by dividing the induced signal u^θ by the factor q^θ. However, q^θ being the inner product of the receive coil sensitivity p and the unit vector in direction of the total magnetic field $\hat{n}_\xi$ as introduced in Eq. (7.25) is equal to zero if the direction of the FFL movement is perpendicular to the direction of the receive coil sensitivity. At this specific angle the induced signal is zero independent of the particle concentration and therefore it is not possible to resolve the whole image and division by p^θ is prohibited.

As a result, reconstruction of the complete Radon data of the SPIO particle distribution requires signal from more than one receive coil. Assuming a system including two receive coils with sensitivities p_1 and p_2 oriented along different directions the voltage signal induced in these receive coils is given by

$$u^\theta(t) = u_1^\theta(t) \pm u_2^\theta(t) = \left(q_1^\theta \pm q_2^\theta\right) A\Lambda'(t) \; (\tilde{\mu} * \mathcal{R}(c)(\theta, \cdot)) \left(\frac{A}{g}\Lambda(t)\right), \tag{7.35}$$

where equivalent to Eq. (7.25), two scalar factors

$$q_1^\theta := p_1 \hat{n}_\xi \tag{7.36}$$

and

$$q_2^\theta := p_2 \hat{n}_\xi \tag{7.37}$$

are defined.

Normalizing this voltage signal with respect to the receive coil sensitivities hence corresponds to division by the factor $(q_1^\theta \pm q_2^\theta)$

$$\tilde{u}^\theta(t) := \frac{u^\theta(t)}{q_1^\theta \pm q_2^\theta} = A\Lambda'(t) \; (\tilde{\mu} * \mathcal{R}(c)(\theta, \cdot)) \left(\frac{A}{g}\Lambda(t)\right). \tag{7.38}$$

To prohibit division by zero, the sign of the signal summation is chosen to prevent signal cancellation. For distinct direction of the receive coil sensitivities p_1 and p_2, these vectors linearly independent. In turn, q_1^θ and q_2^θ are never zero simultaneously.

7.1.4.2 FFL Speed Normalization

In a next step, the signal needs to be compensated for the varying FFL speed. Therefore, the signal $\tilde{u}^\theta(t)$ is normalized with respect to the derivative of the excitation function

$$s^\theta(t) := \frac{\tilde{u}^\theta(t)}{A\Lambda'(t)} = (\tilde{\mu} * \mathcal{R}(c)(\theta, \cdot))\left(\frac{A}{g}\Lambda(t)\right) \tag{7.39}$$

Again, division by zero needs to be prevented. In the case of maximum FFL displacement, the FFL speed is equal to zero. Hence, the signal at this position needs to be neglected.

7.1.4.3 Gridding on Spatial Interval

Another problem occurs due to the fact that the signal is measured at equidistant temporal positions. Due to the varying FFL speed, however, this leads to non-equidistant spatial positions. To evaluate the convolution an interpolation of the signal is needed and the signal is gridded onto equidistant spatial positions. To do so, a transformation of the time signal $s^\theta(t) : \left(0, \frac{T}{2}\right) \to \mathbb{R}$ into a signal depending on the spatial position of the FFL $\tilde{s}^\theta(\xi_{\text{FFL}}) : \left(-\frac{A}{g}, \frac{A}{g}\right) \to \mathbb{R}$ is needed. The coordinate transform

$$\xi_{\text{FFL}}(t) = \frac{A}{g}\Lambda(t)$$

$$t = \Lambda^{-1}\left(\frac{g}{A}\xi_{\text{FFL}}\right) \tag{7.40}$$

is introduced and the transformed signal is hence given by

$$\tilde{s}^\theta(\xi_{\text{FFL}}) := (\tilde{\mu} * \mathcal{R}(c)(\theta, \cdot))(\xi_{\text{FFL}}) \tag{7.41}$$

7.1.4.4 Deconvolution

Finally, to reconstruct the image, deconvolution of the transformed MPI signal is performed. To reconstruct the image, the convolution in Eq. (7.41) has to be inverted. Since a convolution in space corresponds to a multiplication in Fourier space,

$$\hat{s}^{\theta}(v) = \hat{\mu}(v)\,\hat{R}(c)(\theta, v) \tag{7.42}$$

is fulfilled, where $\hat{s}^{\theta}(v) := \mathcal{F}(\tilde{s}^{\theta}(\xi))$, $\hat{R}(c)(\theta, v) := \mathcal{F}(\mathcal{R}(c)(\theta, v))$, and $\hat{\mu}(v) := \mathcal{F}(\tilde{\mu}(\xi))$. The Radon transform of the concentration of the SPIO particle distribution is hence given by

$$\mathcal{R}(c)(\theta, \xi) = \mathcal{F}^{-1}\left(\frac{\hat{s}^{\theta}(v)}{\hat{\mu}(v)}\right). \tag{7.43}$$

Unfortunately, an exact deconvolution is only possible for noiseless data and a kernel, which has non-zero values of its Fourier transform for all frequencies. For a real setup, this will never be provided. The Wiener deconvolution [BM03, KES+11] applies a filter in frequency space damping components of low SNR

$$\tilde{\mathcal{R}}(c)(\theta, \xi) = \mathcal{F}^{-1}\left(\frac{\hat{s}^{\theta}(v)}{\hat{\mu}(v)}\left(\frac{|\hat{\mu}(v)|^2}{|\hat{\mu}(v)|^2 + \frac{1}{SNR(v)}}\right)\right). \tag{7.44}$$

7.2 A Simulation Study

In [KES+11], general feasibility of efficient Radon-based reconstruction in FFL imaging in MPI is demonstrated. The results presented in [KES+11] correspond to approximately ideal conditions. A scanner setup consisting of a ring of four equally sized Maxwell coil pairs generating the FFL selection field of $2.5\,\text{Tm}^{-1}\mu_0^{-1}$ gradient strength as well as two additional Helmholtz coil pairs generating the FFL drive field is assumed. Such a setup is illustrated in Fig. 7.3. In [KES+11], the bore diameter of the scanner is set to 0.30 m and the drive field amplitude amounts to $20\,\text{mT}\mu_0^{-1}$ resulting in an FFL displacement of -8 mm to 8 mm. An excitation frequency of 25 kHz resulting in a period of $T = 40\,\mu s$ is applied and the FFL trajectory consists of 80 discrete, equidistant angles between 0 and π leading to an acquisition time of 3.2 ms.

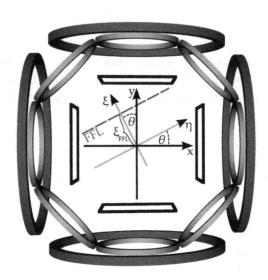

Figure 7.3 Illustration of the FFL scanner setup used in the presented simulation study.

The receive coils used in [KES+11] are equal to those presented in [WBG07, WGB08], which are perpendicularly orientated and located 0.10 m away from the center of the scanner setup (Fig. 7.3). For these coils, the noise level induced by a patient is known [WBG07, KES+11].

In the presented simulation study the results achieved in [KES+11] are expanded by applying FBP on the Radon data derived in the previous section 7.1 for different ratios of the bore diameter of the scanner and the size of the FOV to represent a more realistic situation. These results are compared to images reconstructed using conventional system function based reconstruction to analyze artifacts arising for non-ideal field shapes using Radon-based reconstruction algorithms.

7.2.1 Scanner Setup and Phantom

The design of the scanner setup for the presented simulation study is chosen similar to the design used in [KES+11] and is illustrated in Fig. 7.3. The phantom covers an FOV of 40×40 mm² and consists of SPIO particles of 30 nm in diameter forming the letters IMT as shown in Fig. 7.4. The phantom depth in z-direction, i.e. in direction of the scanner axis, amounts to 10 mm. The size and shape of the phantom is fixed for all presented reconstructions.

Figure 7.4 The IMT phantom with a size of 40×40 mm^2.

The quality of the magnetic fields concerning field homogeneity along the FFL as well as parallel to its alignment strongly depends on the position of interest within the FOV. The field homogeneity is higher close to the center of the scanner and decreases approaching the coil setup. Choosing the bore diameter of the scanner to be preferable large for a given FOV size will hence enhance the magnetic field quality. To analyze the influence of magnetic field inhomogeneities on the image quality achieved using efficient Radon-based reconstruction algorithms, reconstructed results of the introduced phantom using scanner setups of three different bore diameters are presented. A bore of 0.22 m diameter corresponds to a selection field assembly tightly enclosing the receive coils. A bore of 0.30 m diameter equals the setup used in [KES⁺11]. However, a larger FOV than in [KES⁺11] is considered to resemble more realistic conditions.

Table 7.1 Properties of the scanner setup used for the simulation study on efficient Radon-based reconstruction.

Property	Value		
Gradient stregth	$2.5 \text{ Tm}^{-1}\mu_0^{-1}$		
Field of view size	40×40 mm^2		
Drive field amplitude	$60 \text{ mT}\mu_0^{-1}$		
	Setup 1	Setup 2	Setup 3
Bore diameter	0.22 m	0.30 m	2.00 m

Finally, a 2.00 m bore diameter is tested corresponding to an almost ideal scanner. Besides the bore diameter and the drive field amplitude, the properties of the scanner setups are equivalent to [KES+11]. As in [KES+11], the receive coil assembly corresponds to the one introduced in [WBG07] and is not varied for the different scanner setups. The properties of the three considered scanner assemblies are summarized in Tab. 7.1. In the simulations, three different SPIO particle concentration are be assumed, 100 mmol l^{-1}, 10 mmol l^{-1}, and 1 mmol l^{-1} and the SPIO particle core diameter is set to 30 nm.

7.2.2 Results - Signal Transformation

As discussed in section 7.1.4, signal transformations are needed to obtain the data in Radon space. These transformations can be divided in four steps

Step 1 receive channel normalization,

Step 2 FFL speed normalization,

Step 3 gridding on a spatial interval,

Step 4 and deconvolution.

In Fig. 7.5, the signal corresponding to each one of these steps is illustrated. The signal separately induced in both receive coils is shown in Fig. 7.5(a) and Fig. 7.5(b). Fig. 7.5(c) corresponds to the superposition of these signals normalized with respect to the receive sensitivities (Eq. (7.38)). The FFL speed normalized signal according to Eq. (7.39) is illustrated in Fig. 7.5(d). In the next step, the MPI signal is gridded onto a spatial interval (Eq. (7.39)) shown in Fig. 7.5(e). Finally, a Wiener deconvolution is performed according to Eq. (7.44) (Fig. 7.5(f)). This data is then reconstructed using Matlab's (MathWorks) standard FBP. The reconstructed results are presented in the next section.

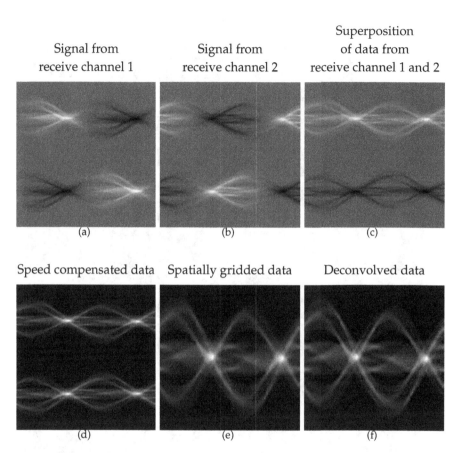

Figure 7.5 MPI data corresponding to the four steps of signal transformation to obtain data in Radon space. (a) and (b) show the voltage signal induced in two perpendicular receive coils. The superposition of these signals normalized with respect to the receive coil sensitivities is illustrated in (c) corresponding to the first step of signal transformation. In the second step, the data is compensated for the FFL velocity with the result shown in (d). Spatial gridding is performed in step three (e) and finally the data is deconvolved in step four (f). Here, the data for the highest SPIO particle concentration and the scanner of 0.30 m bore diameter is shown.

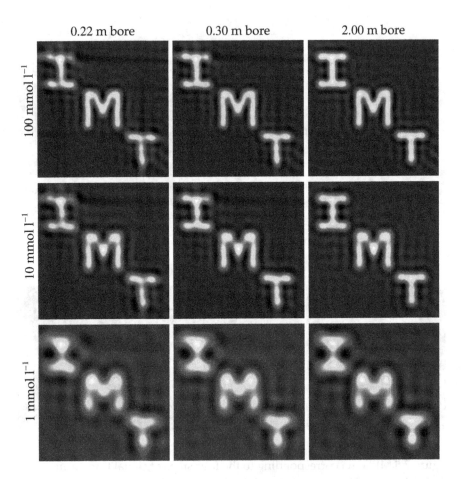

Figure 7.6 Radon-based reconstructed images for SPIO particle concentrations of 100 mmol l^{-1} (upper row), 10 mmol l^{-1} (center row), and 1 mmol l^{-1} (lower row). The left column shows the results for a scanner of 0.22 m bore diameter, the center column for 0.30 m bore diameter and the left column for an ideal scanner.

7.2.3 Results - Reconstructed Images

In the mathematical derivation presented in section 7.1, ideal magnetic fields are assumed. This is certainly only valid in a very small area in the center of the scanner setup. As discussed in chapter 6, a loss in field quality is observed with increasing distance to the scanner center. To analyze the influence of the decreasing field quality, the presented simulation study is performed for three different scanner bore diameters.

The reconstructed results are illustrated in Fig. 7.6. The left column shows the results for the scanner of 0.22 m bore diameter, the center images are obtained for the 0.30 m bore, and the images on the right result from the almost ideal scanner of 2.00 m in diameter. In these images, the influence of the quality of the magnetic fields on the image quality achieved using efficient Radon-based reconstruction is demonstrated.

The image quality is worst regarding the images resulting from the smallest scanner of 0.22 m bore diameter. The M in the center of the images shows nearly no artifacts, which results from a high field homogeneity long the FFL and parallel to its alignment in a region around the center of the scanner assembly. Regarding regions of higher distance to the center, artifacts arise in the image. The I and the T are both distorted and streak artifacts can be observed especially in direction of the upper line of the I. These artifacts are slightly reduced using a scanner of 0.30 m in diameter as illustrated in the center row of Fig. 7.6. The streak artifacts are still visible, but reduced and the distortion is also decreased. Regarding the fact that these results are achieved with simulation parameters according to [KES+11] except for the FOV, which is increased by a factor of 8.2, the influence of the relation between FOV size and bore diameter of the scanner is demonstrated. The results in [KES+11] show almost no artifacts, whereas in the presented results of a more realistic scenario, artifacts arise. It is hence necessary to either increase the magnetic field quality or to develop artifact reduction algorithms.

The enhancement in magnetic field quality was shown in the previous section 6 and these results will be transferred to efficient Radon-based reconstruction in section 7.2.4. In the images on the right side of Fig. 7.6 resulting from an almost ideal scanner setup, almost no streak artifacts and also no distortion is visible.

The presented analysis demonstrates the connection between magnetic field quality and image quality achieved when performing efficient Radon-based reconstruction in FFL imaging in MPI.

7.2.4 Influence of Magnetic Field Quality on Image Quality Achieved in Radon-based FFL Reconstruction

As discussed in chapter 6, the magnetic field quality of a scanner setup consisting of curved rectangular coils is superior to that of all formerly introduced designs. The question arising is how magnetic field quality influences the image quality achieved using the Radon-based reconstruction methods introduced in this chapter. To answer this question, a simulation study using a scanner composed of curved rectangular coils is performed and the results are compared to the results achieved in the last chapter.

The images in the left column of Fig. 7.7 show results obtained with the design utilized in the last chapter with a bore diameter of 0.22 m, while in the right column the results for and a design consisting of curved rectangular coils with identical parameters, i.e. bore diameter, gradient strength, and reconstruction parameters are illustrated.

These results demonstrate that scanner optimization considerations have a positive influence on the image quality achieved using Radon-based reconstruction algorithms. The streak artifacts along the direction of the upper line of the I are reduced when using a scanner of curved rectangular coils. Even more obvious is the reduction of the distortion of the I and the T resulting from the field homogeneity along the FFL in a larger region around the center achieved when using curved rectangular coils.

The presented results emphasize the significance of magnetic field quality optimization when aiming on utilizing efficient Radon-based reconstruction algorithms in FFL imaging in MPI.

7.3 Summary and Discussion

Three major aspects have been presented in this chapter. First, it has been mathematically proven that MPI data acquired using an FFL can be transformed into Radon space. Secondly, it has been demonstrated that artifacts arise for a realistic ratio between the FOV size and the bore diameter of the scanner. And third, it has been shown that these artifacts are reduced when optimizing the magnetic field quality.

Artifact reduction in efficient Radon-based FFL reconstruction will be an important topic for future research. Artifacts arise due to the fact that in the mathematical derivation presented in section 7.1, constituting the fundamentals for the performed reconstructions, data acquisition along a straight line is assumed. This is not fulfilled in a realistic scanner setup and hence leads to the formation of artifacts. There are two ways of reducing the discrepancy between theory and experiment, either the experiment can be optimized to resemble theory as much as possible, or the theory needs to be adjusted to include realistic field shapes. An FFL scanner design providing high magnetic field quality is presented in chapter 6, and as demonstrated in section 7.2.4 artifact reduction is achieved with this design of enhanced magnetic field quality. However, design optimization will always be limited when taking into account practical implementation of the scanner setup. The manufacturing of complicated coil designs is a challenge and electrical power consumption always has to be kept in mind. Hence, a combination of optimizing the experiment and including field inhomogeneities into imaging theory might be a solution to reduce artifact formation in Radon-based FFL reconstruction. In future work, the realistic shape of the magnetic fields results from an optimized FFL imager should be included into the mathematical model.

In the presented simulations, Matlab's standard FBP is used to reconstruct the Radon-data. More advanced reconstruction methods could be applied to enhance image quality.

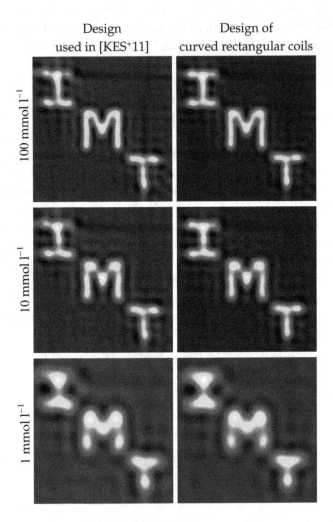

Figure 7.7 Comparison of Radon-based reconstructed images using the design illustrated in Fig. 7.3 and used for the simulation study presented in the last section 7.2.3 and in [KES$^+$11] (left row) with the results achieved using a design of curved rectangular coils (right row). The SPIO particle concentration varies from 100 mmol l^{-1} (upper row) to 10 mmol l^{-1} (center row), and 1 mmol l^{-1} (lower row).

Installation of a Dynamic Field Free Line Imaging Device

In this chapter, the implementation of the first dynamic FFL mouse scanner is presented. According to the findings of chapter 6, the scanner is built of curved rectangular coils and does hence provide an experimental validation for the simulated results of chapter 6. Fig. 8.1(a) shows a picture of the complete transmit coil assembly, i. e. selection and drive field coils, which are all of curved rectangular shape. The complete scanner with the transmit coil assembly mounted on a patient tube and hosted inside of an acrylic glass tube to ensure air tightening is illustrated in Fig. 8.1(b). Air cooling is achieved by the use of six radial fans. The scanner is optimized and implemented to generate a gradient strength of $1.5 \text{ Tm}^{-1}\mu_0^{-1}$. With the presented MPI scanner it will be possible to perform the first dynamic FFL imaging experiments.

8.1 Materials and Methods

An experimental setup is presented, which allows for validation of the simulated results on the optimization of the magnetic field quality as well as the electrical power consumption of an FFL imaging system presented in chapter 6 and [ESKB12]. Curved rectangular coils are used for the implementation of a dynamic FFL scanner setup for the first time. With the presented setup, it will be possible to perform the first dynamic FFL imaging experiments.

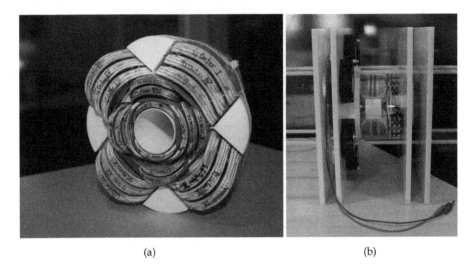

(a)   (b)

Figure 8.1 Implemented transmit coil assembly (a) and the first dynamic FFL scanner mounted in a body ensuring appropriate cooling (b) [Web12].

To validate the manufacturing process of the selection field as well as the drive field coils, the magnetic field quality of the generated fields needs to be determined. The presented results give evidence to the correct implementation of the field generating electromagnetic coils and permanent magnets.

After specifying the scanner properties and describing the construction of the scanner, the transmit coil assembly as well as the complete transmit chain will be presented. To completely describe the dynamic FFL scanner, the receive coil assembly as well as the receive chain are subsequently introduced.

8.1.1 Scanner Properties

The presented dynamic FFL imager is constructed for mouse experiments. The transmit coil assembly illustrated in Fig. 8.1(a) has an internal bore diameter of 40 mm. It is mounted on a CFK tube of 38 mm external diameter to ensure air cooling between the coils and the tube. The CFK tube will accommodate the receive coil assembly shown in Fig. 8.7, which has an internal diameter of 30 mm. The mouse will be placed inside of this receive coil assembly. The

FFL scanner is optimized for a gradient strength of 1.5 $Tm^{-1}\mu_0^{-1}$. This gradient was chosen to provide the possibility to apply an air cooling concept. With the presented scanner, a drive field amplitude of 16 mT will be used to scan an FOV of 1.5×1.5 mm^2.

For the first time, permanent magnets are used to generate the static part of the FFL selection field current. The permanent magnets replace the z-coils introduced in section 3.3. This considerably lowers the overall power consumption of the scanner setup. With this alteration, the concept of splitting the FFL selection field current into a static and a dynamic part enabling the use of permanent magnets is exploited for the first time. Since the field generators of the static selection field current were a crucial aspect concerning electrical power consumption in formerly presented setups, this is a major improvement.

8.1.2 Construction

As the FFL field demonstrator presented in chapter 5, the dynamic FFL imager is constructed using the 3D CAD program SolidWorks (Dassault Systèmes).

This construction program is used to plan the following concepts:

1. the animal tube hosting the receive coil assembly as well as the mouse,

2. the positioning of the transmit coils on the animal tube as well as with respect to each other,

3. the acrylic glass tube outside of the transmit coils ensuring air tightening,

4. the positioning of animal tube and the tube outside of the transmit coil assembly between two POM panels, which ensures mechanical stability of the setup,

5. the cooling concept using six radial fans (EBM PAPST - RL48-19/12 [epSGGCK]), and

6. the positioning and copper shielding of the permanent magnets, which are fixed via two additional POM panels.

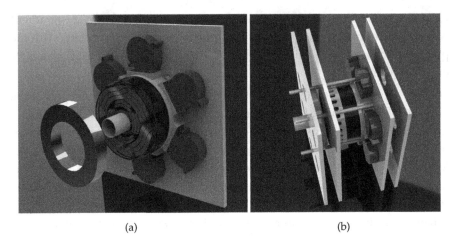

(a)   (b)

Figure 8.2 (a) Construction image of the patent tube, the transmit coils assembly mounted on it, and the acrylic glass tube surrounding the transmit coil assembly. The six fans are installed on one POM panel and one of the permanet magnets is visible. The remaining components are not shown due to facility of inspection. (b) Construction image of the complete FFL scanner housing. Construction realized within the Master's thesis of M. Weber [Web12].

Construction images of the transmit coils assembly, the permanent magnets, and the scanner housing including the cooling concept are illustrated in Fig. 8.2. The construction as well as the implementation of the transmit coil assembly presented in the subsequent section 8.1.3 have been a part of the master's thesis of M. Weber [Web12].

8.1.3 Transmit Coil Assembly

The presented scanner is built of curved rectangular coils. The manufacturing of these coils is more complicated and error-prone than the implementation of circular coils. Before deciding to use such coils for the implementation of an FFL scanner, a test assembly was manufactured and evaluated (section 6.1.6). Only after the results proofed a high agreement with simulated data, the decision to use curved rectangular was made. To manufacture the coils, pressing forms made of steal and POM are constructed.

The coils are winded into the form, glued, and pressed to ensure shape stability. In Fig. 8.3 an examples for a manufactured result is shown.

The transmit coil assembly can be split into selection field coil pairs (SFCPs) and drive field coil pairs (DFCPs). To generate a complete dynamic FFL trajectory (section 3.2), four SFCPs and two DFCPs are needed. Each SFCP is manufactured with a defined radius of curvature $r_{SF,i}$ and each coil covers an opening angle of 90 degrees. It is therefore possible to mount two identically manufactured SFCP on a tube of radius $r_{SF,i}$, covering the complete circumference of the tube (Fig. 8.4). Hence, the four SFCPs can be mounted on an inner and an outer circle. The corresponding coils will be named inner and outer SFCP in the following.

The DFCPs are also manufactured with a defined radius of curvature $r_{DF,i}$. However, they cover a larger opening angle of approximately 120 degrees, which is a preferred choice for generating a homogeneous magnetic field [GM70]. It is therefore not possible to install both DFCPs on a tube with fixed radius. The two DFCPs need to be mounted on top of one another, illustrated in Fig. 8.4, introducing an inner and an outer DFCP.

To achieve gradient strength of $1.5 \, \mathrm{Tm^{-1}} \mu_0^{-1}$, appropriate air cooling is necessary. Splitting the coils into layers increases the surface of the coils, where the air cooling takes effect. The inner DFCP is located directly on the patient tube with the outer DFCP mounted on it. Both DFCPs are hence very close to the FOV. Therefore, their electrical power consumption and current density is not critical. They are manufactured of two litz wire layers, which are not separated.

Figure 8.3 Implemented curved rectangular coils.

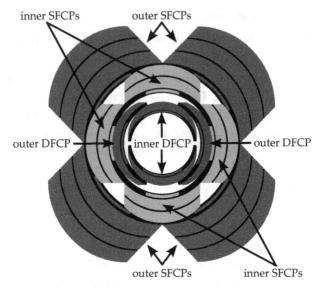

Figure 8.4 SFCP and DFCP.

The inner and the outer DFCP are separated with a distance of 1.5 mm, so cooling of the whole surface of each coil is guaranteed. The inner SFCPs are separated into two layers with a distance of 1.5 mm, while the outer SFCPs consist of three layers.

The complete transmit coil assembly is illustrated in Fig. 8.5. On the left, a construction image is illustrated, whereas on the right photo of the implemented coils is shown. The inner two coil pairs are used for drive field generation, while a combination of the remaining coil pairs generates the dynamic part of the selection field. The presented electromagnetic coils need to be accompanied by a set of permanent magnets, which generate the static part of the selection field.

8.1.4 Transmit Chain

The transmit chain of the presented FFL mouse scanner consists of the following components:

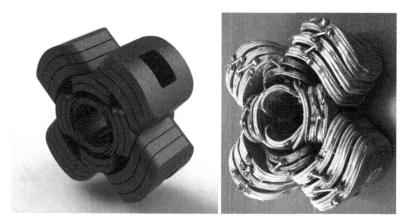

Figure 8.5 On the left, a constructed image of the transmit coil assembly generating the drive field (DFCP) as well as the dynamic part of the selection field (SFCP) is shown. The right image shows a photo of the implemented coil setup [Web12].

1. PC including digital-to-analog converter and DC source control,

2. DC source (Delta Elektronika SM 7.5-80 [Ele]) to apply current to the selection field coils. In the presented scanner, a discrete FFL trajectory is realized. Hence, the selection field currents are applied by DC sources. To realize a continuous trajectory, the selection field will be realized via AC current.

3. AC power amplifier (AE Techtron 7796 [Sha]),

4. band-stop filter to ensure that the excitation frequency of the drive field does not couple into the DC source through the selection field coils,

5. 3rd order Butterworth band-pass filter to ensure a pure sinusoidal excitation signal, and

6. transmit coil assembly, which has been discussed in the previous section 8.1.3.

The transmit chain of the presented FFL scanner illustrated in Fig. 8.6 as well as the receive chain presented in section 8.1.6 were implemented as part of the master's thesis of K. Bente [Ben13].

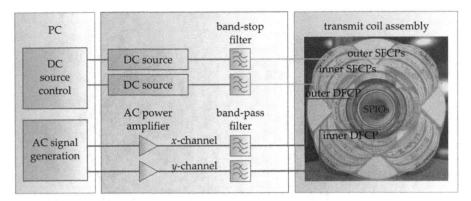

Figure 8.6 Transmit chain of the implemented dynamic FFL mouse scanner.

8.1.5 Receive Coil Assembly

The receive coil assembly detects the voltage signal induced due to the change of the SPIO particle magnetization when exposed to the MPI drive field. It is optimized to provide a high magnetic field homogeneity. It was implemented as part of the bachelor's thesis of S. Heinitz [Hei12]. It consists of two coil pairs with axes in x- and y-direction. The receive coils are manufactured of litz wire with a diameter of x mm and a photo of the coil pairs is shown in Fig. 8.7. The distance of the single windings corresponds to a sine function. The coils are glued to ensure stability. The receive coil pairs in x- and y-direction are mounted on one another.

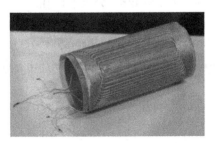

Figure 8.7 Receive coil assembly. Two receive coil pairs are fit into each other, rotated by 90 degree with respect to one another.

8.1.6 Receive Chain

The receive chain of the presented FFL scanner is illustrated in Fig. 8.8 and consists of the following components:

1. the receive coil assembly introduced in the previous section 8.1.5, which detects a voltage signal induced due to the change of the SPIO particles magnetization,

2. band-stop filter to damp the excitation signal coupling into the receive coil,

3. a low noise amplifier to amplify the SPIO particle signal,

4. an analog-to-digital converter, and

5. the hard disc of the PC, where the measured data is saved and processed.

All of the presented components work together to provide the technical instrument needed to achieve dynamic FFL imaging in MPI.

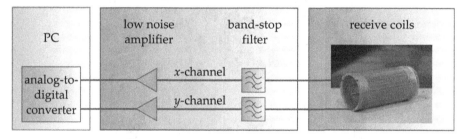

Figure 8.8 Receive chain.

8.1.7 Magnetic Field Measurements.

To evaluate the quality of the experimental setup, a rotated and translated FFL field emulating the complete dynamic FFL trajectory is generated, measured and compared to simulated results. To validate the magnetic field quality, it is not necessary to generate the final gradient strength of $1.5 \text{ Tm}^{-1}\mu_0^{-1}$. A smaller gradient of $0.4 \text{ Tm}^{-1}\mu_0^{-1}$ is therefore chosen for the evaluation. The resulting fields are measured on a field of view (FOV) of $10 \times 10 \text{ mm}^2$ and 25×25 pixel. To evaluate the magnetic field quality, the measured results are compared to simulated fields, which were predetermined using numerical evaluation of the Biot-Savart integral [Jac99], in terms of the normalized root mean square deviation (NRMSD).

The generated fields were measured with a Gauss meter (LakeShore Model 460) with a three dimensional Hall sensor. The sensor was moved through the FOV using a robot (Iselautomation GmbH & Co. KG).

Figure 8.9 The dynamic FFL scanner setup. On the right, the FFL scanner consisting of transmit and receive coil assembly installed in the housing is mounted on a table. The robot will be used to determine the measurement-based system function. The filtering components are located beneath the table [Web12].

Table 8.1 Currents applied to the inner and outer SFCPs to realize an FFL field rotation.

FFL rotation angle / degree	0	22.5	45	67.5	90	112.5	135	157.5
Current inner SFCP / A	9.5	6.7	0	6.7	9.5	6.7	0	6.7
Current outer SFCP / A	0	15.1	21.3	15.1	0	15.1	21.3	15.1

8.2 Results

A dynamic FFL scanner setup, including transmit chain with curved rectangular transmit coil assembly and the complete receive chain is presented. Fig. 8.9 shows a photo of the scanner mounted on a table, the robot, which will be used to determine the measurement based system function as well as to host the mouse bed, and the filter setup beneath the table. The presented FFL scanner is capable of generating a complete dynamic FFL trajectory.

The generated gradient strength of $0.4\ \mathrm{Tm}^{-1}\mu_0^{-1}$ is considerably higher than the gradient strength achieved in former measurements using circular coils [EKS+11a]. Although the gradient strength is increased, the total power consumption of the setup is reduced to 37 W, compared to 42 W, which was needed for the generation of a $0.25\ \mathrm{Tm}^{-1}\mu_0^{-1}$ gradient with a setup of circular coils in [EKS+11a]. The results provide an experimental proof for the feasibility of the simulated results presented in chapter 6 and [ESKB12].

8.2.1 FFL Field Rotation

A magnetic FFL field rotated between 0 and 157.5 degree is generated and measured at 8 equidistant FFL rotation angles. Since a 180 degree rotation corresponds to the FFL field of 0 degree rotation, it is not measured. The currents applied to the inner and outer SFCPs are specified in Tab. 8.1 for each rotation angel.

The measured data of the FFL field rotation is illustrated in Fig. 8.10. The FFL is rotated by applying the appropriate currents to the selection field coils only, no mechanical movement is involved.

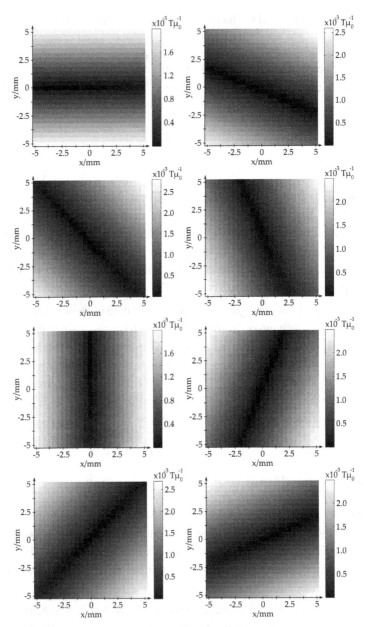

Figure 8.10 Measured results of the FFL field rotation. 8 distinct FFL rotation angles are realized.

Table 8.2 NRMSD of the rotated FFL field at 8 equidistant FFL angles compared to simulated data.

Rotation angle / degree	0	22.5	45	67.5	90	112.5	135	157.5
NRMSD / %		1.55	4.41	3.53	2.72	1.91	1.86	2.05	1.11

Table 8.3 NRMSD of the FFL field compared to simulated results for vertical, diagonal, and horizontal translation.

Orientation of the FFL translation	vertical	diagonal	horizontal
NRMSD / %	2.42	1.44	4.42

The NRMSD is given in Tab. 8.2. It varies between 1.11 % and 4.41 % and is therefore comparable to the NRMSD achieved in the assembly of circular coils [EKS+11a]. Hence, no additional errors were introduced due to manufacturing inaccuracy. These results constitute an experimental proof for the feasibility of the simulation study identifying curved rectangular coils as the optimal choice for magnetic field generation in dynamic FFL imaging in MPI.

8.2.2 FFL Field Translation

In addition to the FFL rotation, the translation of the FFL field realized via the DFCPs is needed for an imaging experiment.

Therefore, the DFCPs are used to achieve a constant translation of the FFL field, which is measured and evaluated in terms of the NRMSD with respect to simulated data. The measured results of a horizontal, diagonal as well as vertical FFL field translation are illustrated in Fig. 8.11. Tab. 8.3 provides the NRMSD for the translation data. The field errors occurring for the simultaneous use of SFCPs and DFCPs during the translation process do not differ from the errors for the pure rotation, which only involved the SFCPs. Hence, the results achieved after implementation of the DFCPs also agree with the simulations.

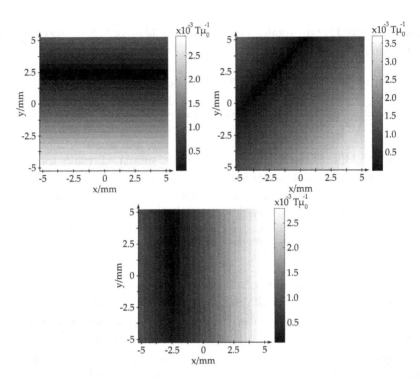

Figure 8.11 Measured results a combined FFL field rotation and translation. Vertical, diagonal, and horizontal FFL field translation has been performed.

8.2.3 Electrical Power Consumption

Not only the magnetic field quality, but also the electrical power consumption of each coil is measured and compared to the results expected from simulations. This is separately done for each selection and drive field coil. Since the electrical power consumption is measured, the two coils constituting a SFCP or DFCP are separately considered. In the case of the selection field coils, the rotation angle of maximum power consumption of each coil type is considered. Hence, the resulting 37 W in total constitute the maximum value.

The measured and simulated electrical power consumption of the four inner selection field coils are compared in Tab. 8.4. The same is done for the three combined outer selection field coils. All four of those coupled coils were also

Table 8.4 Comparison of the measured and simulated results of the electrical power loss in the inner selection field coils.

Coil number	voltage in V (measured)	power loss in W (measured)	power loss in W (simulated)
1	0.196	1.86	1.59
2	0.162	1.53	1.59
3	0.160	1.52	1.59
4	0.157	1.50	1.59

Table 8.5 Comparison of the measured and simulated results of the electrical power loss in the outer selection field coils.

Coil number	voltage in V (measured)	power loss in W (measured)	power loss in W (simulated)
1	0.368	7.84	6.54
2	0.352	7.51	6.54
3	0.388	8.27	6.54
4	0.345	7.34	6.54

measured and the results for the voltage and the simulated and the measured electrical power consumption are given in Tab. 8.5. The power consumption of the drive field coils is given in Tab. 8.6.

The results agree to a very high extend with the predicted simulation data. Only a small rise in electrical power consumption resulting from implementation effects is observed.

8.3 Summary and Discussion

Simulation studies are used to predict and optimize the performance of MPI imaging systems. However, the implementation of a specific MPI scanner, although simulated in detail, might hold challenges, which are not taken into account in the simulation. It is therefore of great importance to validate simu-

Table 8.6 Comparison of the measured and simulated results of the electrical power loss in the drive field coils.

Coil number	voltage in V (measured)	power loss in W (measured)	power loss in W (simulated)
1	0.021	0.06	0.06
2	0.022	0.06	0.06
3	0.035	0.15	0.15
4	0.035	0.15	0.15

lated scanner designs with respect to their experimental feasibility. This is even more important, when complicated coil shapes like curved rectangular coils are involved, as in the presented study. It is for example not possible to implement purely rectangular coils, since the litz wire used for coil production has a certain radius of curvature. This was not taken into account in prior simulations and might have led to errors in the resulting magnetic fields. Fortunately, the presented results show that the measured field errors are comparable to those occurring in a setup of circular coils [EKS+11a].

With the presented setup, a complete dynamic FFL trajectory of 0.4 $Tm^{-1}\mu_0^{-1}$ gradient strength was generated with a maximum power consumption of 37 W. This outperforms the experimental setup of circular coils presented in [EKS+11a], since for a gradient strength of only 0.25 $Tm^{-1}\mu_0^{-1}$ an electrical power consumption of 42 W was needed. Hence, a simultaneous improvement in gradient strength and electrical power consumption is achieved.

The presented experimental validation study provides a proof of experimental feasibility for the magnetic field generation for a complete dynamic FFL trajectory using an efficient setup of curved rectangular coils. The presented scanner setup constitutes the foundation for an efficient and precise dynamic FFL imaging system. Each component of the transmit and receive chain is installed and tested. The simultaneous initiation of all components is to be realized in a next step to perform dynamic FFL imaging with the presented setup. An disturbance frequency of 4 MHz prohibits imaging experiments at the moment and needs to be compensated for. After extinguishing this error source, a complete setup for dynamic FFL imaging in MPI is provided.

CHAPTER 9

Summary

The presented thesis constitutes a first step towards the vision of clinical application of dynamic FFL imaging in MPI. Various aspects of FFL imaging in MPI are examined and some very promising research fields like efficient Radon-based reconstruction and dynamic FFL scanner instrumentation are initiated.

In the introduction of this thesis, medical applications of MPI are discussed, the magnetic properties of the SPIO tracer particles enabling their detection are described and signal reception is explained. Furthermore, the physical properties enabling signal and spatial encoding are introduced. To completely cover MPI, the system function, as well as MPI reconstruction and the signal chain are discussed. Subsequently, the alternative line encoding scheme analyzed in this thesis is introduced and motivated. The FFL field and trajectory are presented and a description of the different FFL imaging techniques is provided.

A mathematical description of magnetic FFL generation and the electromagnetic components generating these fields is provided and the design proposed in [KEB⁺10] is motivated from a purely mathematical point of view.

To provide an experimental demonstration of feasibility for magnetic field generation for FFL imaging in MPI, the first FFL field demonstrator is implemented, tested, and evaluated within the scope of this thesis. The promising results substantiate preceding simulation studies with a proof of concept and motivate further research on dynamic FFL imaging. The presented FFL demonstrator is capable of generating an FFL field complying with the requirements for dynamic FFL imaging in MPI for the first time.

Prior to this thesis, all proposed FFL scanner designs were composed of circular coils only. An analysis of magnetic field quality and overall electrical power consumption of scanner designs consisting of different coil shapes is presented and curved rectangular coils are identified as the perfect match between these two aspects. It is shown that using curved rectangular coils enhances magnetic field quality by a factor of almost five compared to conventional circular coils, while simultaneously electrical power consumption is reduced by a factor of almost four. A test assembly is implemented to provide an experimental proof of feasibility for the simulated results.

Applying a line detection scheme, the possibility of using efficient Radon-based reconstruction methods arises. A mathematical description of data transformation into Radon space is provided [KES+11] and a simulation study on Radon-based FFL reconstruction is presented. In [KES+11], general feasibility of Radon-based reconstruction in FFL imaging was demonstrated, however, for a small FOV compared to the scanner bore diameter, which resembles almost ideal conditions. In a small region in the center of the scanner setup the magnetic field corresponds to an almost ideal FFL field, while gradient inhomogeneites leading to artifact formation appear approaching the scanner setup. Due to that reason, it is important to consider more realistic relations between FOV size and scanner bore diameter, which is realized in this thesis. Furthermore, the effect of an increased magnetic field quality on the achieved image quality using Radon-based reconstruction algorithms in FFL imaging is analyzed and it is demonstrated that magnetic field quality optimization leads to a reduction of artifact formation [EKSB12].

Finally, the installation of a dynamic FFL mouse scanner is presented. The transmit coil assembly corresponds to the optimized design of curved rectangular coils proposed in this thesis. The magnetic fields generated with the FFL scanner show a high agreement with simulated data. Compared to the results achieved with the FFL field demonstrator [EKS+11a], the FFL scanner generates a higher gradient strength at reduced electrical power consumption and improved magnetic field quality [EWSB13]. The complete transmit and receive chain are installed and presented.

CHAPTER 10

Discussion and Outlook

Since MPI in general, and FFL imaging in particular, is a very young medical imaging technology, fundamental questions concerning for example reconstruction methods, scanner instrumentation, or selection field shape are yet to be answered and have to be investigated in future research.

The analysis of coil shapes presented in this thesis is restricted to shapes generally used in various applications. This restriction originates from the fact that the presented optimization considerations are the preparation of the installation of an FFL imaging device. Therefore, more complicated coil shapes, which could not have been realized within the scope of this work, are neglected. In future research, however, the analysis should be extended to more general conductor assemblies. This will open up new possibilities for magnetic field quality optimization. Where one has to keep in mind that all simulations aiming on field quality optimization should be validated with respect to practical feasibility.

As it is shown in this thesis, gradient inhomogeneities of the FFL field lead to artifact formation using Radon-based reconstruction algorithms. Hence, these reconstruction algorithms should be extended to include the realistic shape of the magnetic fields generated by an FFL imaging device.

In the dynamic FFL mouse scanner presented in this thesis the 4 MHz disturbance frequency needs to be compensated for to acquire the first dynamic FFL images. Furthermore, the presented FFL scanner may not only be used for FFL imaging, but also for FFP imaging. This would enable direct comparison of the sensitivities of FFL and FFP imaging not depending on the technical aspects of the scanner setup. Producing FFP and FFL images with one scanner setup would provide a proof for sensitivity enhancement of the line encoding scheme, which would constitute a major step of establishing the FFL imaging method.

The presented scanner setup is realized with the aim of providing a proof of feasibility for dynamic FFL imaging in MPI. Since it is the world's first setup complying with all requirements imposed by dynamic FFL imaging in MPI, it needs to be considered as a prototype. To achieve a higher gradient strength and a larger FOV the cooling concept needs to be extended to water, oil or liquid nitrogen cooling.

General feasibility of magnetic field generation, efficient Radon-based reconstruction, and major improvements in dynamic FFL scanner design have been presented. High potential lies in the field of dynamic FFL imaging in MPI and will be exploited tying in with the results of this thesis.

Bibliography

[BBB+10] T. M. Buzug, S. Biederer, J. Borgert, M. Erbe, T. Knopp, K. Lüdtke-Buzug, and T. F. Sattel (Eds.). *1st International Workshop on Magnetic Particle Imaging (IWMPI 2010) - Book of Abstracts*. Verlags- und Druckhaus Max Schmidt-Römhild KG, Lübeck, 2010.

[BBE+12] T. M. Buzug, G. Bringout, M. Erbe, K. Gräfe, M. Gräser, M. Grüttner, A. Halkola, T. F. Sattel, W. Tenner, H. Wojtczyk, J. Hägele, F. M. Vogt, J. Barkhausen, and K. Lüdtke-Buzug. Magnetic Particle Imaging: Introduction to Imaging and Hardware Realization. *Zeitschrift für Medizinische Physik*, 22(4):323–334, 2012.

[BBK+10] T. M. Buzug, J. Borgert, T. Knopp, S. Biederer, T. F. Sattel, M. Erbe, and K. Lüdtke-Buzug (Eds.). *Magnetic Nanoparticles: Particle Science, Imaging Technology, and Clinical Applications*. World Scientific Publishing Company, Singapore, 2010.

[Ben13] K. Bente. *Implementation of a Magnetic Particle Imaging System for a Dynamic Field Free Line*. Master's thesis, Institute of Medical Engineering, University of Lübeck, Lübeck, 2013.

[Bie12] S. Biederer. *Magnet-Partikel-Spektrometer: Entwicklung eines Spektrometers zur Analyse superparamagnetischer Eisenoxid-Nanopartikel für Magnetic-Particle-Imaging*. Springer Vieweg, Berlin/Heidelberg, 2012.

[BKS+09] S. Biederer, T. Knopp, T. F. Sattel, K. Lüdtke-Buzug, B. Gleich, J. Weizenecker, J. Borgert, and T. M. Buzug. Magnetization

Response Spectroscopy of Superparamagnetic Nanoparticles for Magnetic Particle Imaging. *Journal of Physics D: Applied Physics*, 42(20):7pp, 2009.

[BKS+10a] S. Biederer, T. Knopp, T. F. Sattel, M. Erbe, and T. M. Buzug. Improved Estimation of the Magnetic Nanoparticle Diameter with a Magnetic Particle Spectrometer and Combined Fields. In *Proceedings of the International Society for Magnetic Resonance in Medicine*, volume 18, page 954, Stockholm, 2010.

[BKS+10b] S. Biederer, S. Kren, T. F. Sattel, M. Erbe, T. Knopp, K. Lüdtke-Buzug, and T. M. Buzug. Ein magnetisches Partikel-Spektrometer zur Messung der Magnetisierung von Nanopartikeln unter der Verwendung von AC- und DC-Feldern. In *Proceedings Biomedizinische Technik / Biomedical Engineering*, volume 55 (Suppl. 1), Rostock, 2010. Walter de Gruyter, Berlin/New York, doi:10.1515/BMT.2010.295.

[BL59] C. P. Bean and J. D. Livingston. Superparamagnetism. *Journal of Applied Physics*, 30:120–129, 1959.

[BM03] H. H. Barrett and K. Myers. Foundations of Image Science. *Wiley, New York*, 2003.

[Bro63] W. F. Brown. Thermal Fluctuations of a Single-Domain Particle. *Physical Review Letters*, 130:1677–1686, 1963.

[BSE+10] T. M. Buzug, T. F. Sattel, M. Erbe, S. Biederer, J. Borgert, D. Finas, K. Dietrich, F. M. Vogt, J. Barkhausen, K. Lüdtke-Buzug, and T. Knopp. Alternative Spulentopologien für Magnetic-Particle-Imaging. *RöFo: Fortschritte auf dem Gebiet der Röntgenstrahlen und bildgebenden Verfahren*, 182:A56, doi:10.1055/s-0030-1268341, 2010.

[BSE+11a] T. M. Buzug, T. F. Sattel, M. Erbe, S. Biederer, D. Finas, K. Diedrich, F. Vogt, J. Barkhausen, J. Borgert, K. Lüdtke-Buzug, and T. Knopp. *Magnetic Particle Imaging: Principles and Clinical Application in Nanomedicine - Basic and Clinical Applications in Diagnostics and Therapy.* pages 88-95, Karger, Berlin, 2011.

[BSE+11b] T. M. Buzug, T. F. Sattel, M. Erbe, S. Biederer, D. Finas, K. Diedrich, F. M. Vogt, J. Barkhausen, K. Lüdtke-Buzug, and T. Knopp. Novel Hardware Developments in Magnetic Particle Imaging. In *Proceedings of the SPIE Symposium on Medical Imaging: Biomedical Applications in Molecular, Structural, and Functional Imaging, 79650T*, pages 1–6, Orlando, doi:10.1117/12.877158, 2011.

[BSK+10a] S. Biederer, T. F. Sattel, T. Knopp, M. Erbe, and T. M. Buzug. Improving the Imaging Quality in Magnetic Particle Imaging by a Traveling Phase Trajectory. In *Proceedings of the International Society for Magnetic Resonance in Medicine*, volume 18, page 3296, Stockholm, 2010.

[BSK+10b] S. Biederer, T. F. Sattel, T. Knopp, M. Erbe, K. Lüdtke-Buzug, F. M. Vogt, J. Barkhausen, and T. M. Buzug. A Spectrometer to Measure the Usability of Nanoparticles for Magnetic Particle Imaging. *Magnetic Nanoparticles: Particle Science, Imaging Technology, and Clinical Applications*, 1:60–65, World Scientific Publishing Company, Singapore, 2010.

[Buz08] T. M. Buzug. *Computed Tomography: From Photon Statistics to Modern Cone-Beam CT*. Springer Verlag, Berlin/Heidelberg, 2008.

[CC64] S. Chikazumi and S. H. Charap. *Physics of Magnetism*. Wiley, New York, 1964.

[Cor63] Cormarck. Representation of a Function by Its Line Integrals, with Some Radiological Applications. *Journal of Applied Physics*, 34(9):2722, 1963.

[Cor64] Cormarck. Representation of a Function by Its Line Integrals, with Some Radiological Applications. II. *Journal of Applied Physics*, 35(10):2908, 1964.

[EGSB12] M. Erbe, M. Grüttner, T. F. Sattel, and T. M. Buzug. Experimentelle Realisierungen einer vollständigen Trajektorie für die magnetische Partikel-Bildgebung mit einer feldfreien Linie. In *Bildverarbeitung für die Medizin*, pages 358–362, Berlin, 2012. Springer Verlag, Berlin/Heidelberg.

[EKB⁺10] M. Erbe, T. Knopp, S. Biederer, T. F. Sattel, and T. M. Buzug. Experimentelle Erzeugung einer magnetischen feldfreien Linie für die Anwendung in Magnetic-Particle-Imaging. In *Proceedings Biomedizinische Technik / Biomedical Engineering*, volume 55 (Suppl. 1), Rostock, 2010. Walter de Gruyter, Berlin/New York, doi:10.1515/BMT.2010.101.

[EKS⁺11a] M. Erbe, T. Knopp, T. F. Sattel, S. Biederer, and T. M. Buzug. Experimental generation of an arbitrarily rotated magnetic field-free line for the use in magnetic particle imaging. *Medical Physics*, 38(9):5200–5207, 2011.

[EKS⁺11b] M. Erbe, T. Knopp, T. F. Sattel, S. Biederer, and T. M. Buzug. Experimentelle Validierung des Konzeptes einer feldfreie Linie für Magnetic-Particle-Imaging anhand von Magnetfeldmessungen. In *Bildverarbeitung für die Medizin*, pages 334–338, Lübeck, 2011. Springer Verlag, Berlin/Heidelberg.

[EKSB12] M. Erbe, T. Knopp, T. F. Sattel, and T. M. Buzug. Influence of Magnetic Field Optimization on Image Quality Achieved for Efficient Radon-Based Reconstruction in Field Free Line Imaging in MPI. *Springer Proceedings in Physics 140*, pages 225–229, 2012.

[Ele] Delta Elektronika. Manual - SM800-series, http://www.schulz-electronic.de/userfiles/file/manual/SM800_B_E.pdf (retrieval date: 23.09.2013).

[epSGGCK] ebm-papst St. Georgen GmbH & Co. KG. RL 48-19/12 ML - DC-Radiallüfter, http://img.ebmpapst.com/products/datasheets/DC-Radialventilator-RL481912ML-GER.pdf (retrieval date: 23.09.2013).

[ESB12a] M. Erbe, T. F. Sattel, and T. M. Buzug. Improved magnetic particle spectrometer providing high field amplitudes for investigation of hysteresis effect in superparamagnetic nanoparticle tracers. In *Proceedings of the World Molecular Imaging Congress*, Dublin, 2012.

[ESB12b] M. Erbe, T. F. Sattel, and T. M. Buzug. Commercialization of a Magnetic Particle Spectrometer. In *Proceedings of the*

IEEE 12th International Conference on Nanotechnology, Birmingham, doi:10.1109/NANO.2012.6321939, 2012.

[ESB13] M. Erbe, T. F. Sattel, and T. M. Buzug. Improved Field Free Line Magnetic Particle Imaging Using Saddle Coils. *Biomedical Engineering*, 10:1–6, doi: 10.1515/bmt-2013-0030, [Epub ahead of print], 2013.

[ESK+11] M. Erbe, T. F. Sattel, T. Knopp, S. Biederer, and T. M. Buzug. An optimized field free line scanning device for magnetic particle imaging. In *Proceedings Biomedizinische Technik / Biomedical Engineering*, volume 56. (Suppl. 1), page BMT.2011.298, Freiburg, 2011. Walter de Gruyter, Berlin/New York, doi:10.1515/BMT.2010.101.

[ESKB12] M. Erbe, T. F. Sattel, T. Knopp, and T. M. Buzug. Enhancing the Efficiency of a Field Free Line Scanning Device for Magnetic Particle Imaging. In *Proceedings of the IEEE Nuclear Science Symposium and Medical Imaging Conference*, pages 2566 – 2568, Anaheim, doi:10.1109/NSSMIC.2012.6551587, 2012.

[EWSB13] M. Erbe, M. Weber, T. F. Sattel, and T. M. Buzug. Experimental Validation of an Assembly of Optimized Curved Rectangular Coils for the use in Dynamic Field Free Line Magnetic Particle Imaging. *Current Medical Imaging Reviews*, 9(2):89–95(7), 2013.

[FRB+10] D. Finas, B. Ruhland, K. Baumann, T. Knopp, T. F. Sattel, S. Biederer, K. Lüdtke-Buzug, T. M. Buzug, and K. Diedrich. Sentinel Lymphnode Detection in Breast Cancer by Magnetic Particle Imaging Using Superparamagnetic Nanoparticles. *Magnetic Nanoparticles: Particle Science, Imaging Technology, and Clinical Applications*, 1:205–210, World Scientific Publishing Company, Singapore, 2010.

[GC11] P. W. Goodwill and S. M. Conolly. Multidimensional X-Space Magnetic Particle Imaging. *IEEE Transactions on Medical Imaging*, 30(9):1581–1590, 2011.

[GGB+11] M. Grüttner, M. Gräser, S. Biederer, T. F. Sattel, H. Wojtczyk, W. Tenner, T. Knopp, B. Gleich, J. Borgert, and T. M. Buzug. 1D-

Image Reconstruction for Magnetic Particle Imaging Using a Hybrid System Function. In *Proceedings of the IEEE Nuclear Science Symposium Medical Imaging Conference*, pages 2545–2548, Valencia, doi:10.1109/NSSMIC.2011.6152687, 2011.

[GKZ⁺12] P. W. Goodwill, J. Konkle, B. Zheng, E. U. Saritas, and S. M. Conolly. Projection X-Space Magnetic Particle Imaging. *IEEE Transactions on Medical Imaging*, 31(5):1076–1085, doi:10.1109/TMI.2012.2185247, 2012.

[GKZC12] P. W. Goodwill, J. Konkle, B. Zheng, and S. M. Conolly. Projection X-Space MPI Mouse Scanner. *Springer Proceedings in Physics 140*, 1:267–274, 2012.

[GM70] D. M. Ginsberg and M. J. Melchner. Optimum Geometry of Saddle Shaped Coils for Generating a Uniform Magnetic Field. *Review of Scientific Instruments*, 41(1):122–123, 1970.

[GSLB⁺12] K. Gräfe, T. F. Sattel, K. Lüdtke-Buzug, D. Finas, J. Borgert, and T. M. Buzug. Magnetic Particle Imaging for Sentinel Lymph Node Biopsy in Breast Cancer. *Springer Proceedings in Physics 140*, pages 237–241, 2012.

[GSV06] D. Gatteschi, R. Sessoli, and J. Villain. *Molecular Nanomagnets*. Oxford University Press, Oxford, 2006.

[Gui09] A. P. Guimaraes. *Principles of Nanomagnetism*. Springer, Berlin/Heidelberg, 2009.

[GvL93] G. H. Golub and C. F. van Loan. *Matrix Computations*. The Johns Hopkins University Press, Baltimore, 2nd edition, 1993.

[GW05] B. Gleich and J. Weizenecker. Tomographic imaging using the nonlinear response of magnetic particles. *Nature*, 435(7046):1214–1217, 2005.

[HBW⁺12] J. Haegele, S. Biederer, H. Wojtczyk, M. Gräser, T. Knopp, T. M. Buzug, J. Barkhausen, and F. M. Vogt. Towards Cardiovascular Interventions Guided by Magnetic Particle Imaging (MPI):

First Instrument Characterization. *Magnetic Resonance in Medicine,* 69(6):1761–1767, doi:10.1002/mrm.24421, 2012.

[Hei12] S. Heinitz. *Realization of a Tube-Shaped Receiving Coil set for Magnetic Particle Imaging with Homogeneous Sensitivity Profiles.* Bachelor's thesis, Institute of Medical Engineering, University of Lübeck, Lübeck, 2012.

[Hou73] G. N. Hounsfield. Computerized transverse axial scanning (tomography). Part I. Description of system. *British Journal of Radiology,* 46:1016–1022, doi:10.1259/0007-1285-46-552-1016, 1973.

[HRG⁺12] J. Haegele, J. Rahmer, B. Gleich, J. Borgert, H. Wojtczyk, N. Panagiotopoulos, T. M. Buzug, J. Barkhausen, and F. M. Vogt. Magnetic Particle Imaging: Visualization of Instruments for Cardiovasculal Intervention. *Radiology,* 265(3):933–938, 2012.

[HSE⁺12] J. Haegele, T. F. Sattel, M. Erbe, K. Lüdtke-Buzug, M. Taupitz, J. Borgert, T. M. Buzug, J. Barkhausen, and F. M. Vogt. Magnetic Particle Imaging (MPI). *RöFo: Fortschritte auf dem Gebiet der Röntgenstrahlen und bildgebenden Verfahren,* 184(5):420–426, doi:10.1055/s-0031-1281981, 2012.

[Jac99] J. D. Jackson. *Classical Electrodynamics.* John Wiley & Sons, New York, 1999.

[Kac37] S. Kaczmarz. Angenäherte Auflösung von Systemen linearer Gleichungen. *Bull. Internat. Acad. Polon. Sci. Lett.,* A35:355–357, 1937.

[Kad59] H. Kaden. *Wirbelströme und Schirmung in der Nachrichtentechnik.* Springer Verlag, Berlin/Heidelberg, 1959.

[KB12] T. Knopp and T. M. Buzug. *Magnetic Particle Imaging: An Introduction to Imaging Principles and Scanner Instrumentation.* Springer Verlag, Berlin/Heidelberg, 2012.

[KBS⁺09] T. Knopp, S. Biederer, T. F. Sattel, J. Weizenecker, B. Gleich, J. Borgert, and T. M. Buzug. Trajectory Analysis for Magnetic Particle Imaging. *Physics in Medicine and Biology,* 54(2):385–397, 2009.

[KBS⁺10a] T. Knopp, S. Biederer, T. F. Sattel, K. Lüdtke-Buzug, M. Erbe, and T. M. Buzug. Efficient Field-Free Line Generation for Magnetic Particle Imaging. *Magnetic Nanoparticles: Particle Science, Imaging Technology, and Clinical Applications*, 1:120–125, World Scientific Publishing Company, Singapore, 2010.

[KBS⁺10b] T. Knopp, S. Biederer, T. F. Sattel, J. Rahmer, J. Weizenecker, B. Gleich, J. Borgert, and T. M. Buzug. 2D model-based reconstruction for magnetic particle imaging. *Medical Physics*, 37:485–491, 2010.

[KBS⁺11a] T. Knopp, S. Biederer, T. F. Sattel, M. Erbe, and T. M. Buzug. Prediction of the Spatial Resolution of Magnetic Particle Imaging Using the Modulation Transfer Function of the Imaging Process. *IEEE Transactions on Medical Imaging*, 30(6):1284–1292, 2011.

[KBS⁺11b] T. Knopp, S. Biederer, T. F. Sattel, M. Erbe, and T. M. Buzug. Über das Auflösungsvermögen von Magnetic Particle Imaging. In *Bildverarbeitung für die Medizin*, pages 329–333, Lübeck, 2011. Springer Verlag, Berlin/Heidelberg.

[KEB⁺10] T. Knopp, M. Erbe, S. Biederer, T. F. Sattel, and T. M. Buzug. Efficient Generation of a Magnetic Field-Free Line. *Medical Physics*, 37(7):3538–3540, 2010.

[KES⁺10a] T. Knopp, M. Erbe, T. F. Sattel, S. Biederer, and T. M. Buzug. Efficient Generation of a Magnetic Field-Free Line. In *Proceedings of the International Society for Magnetic Resonance in Medicine*, volume 18, page 952, Stockholm, 2010.

[KES⁺10b] T. Knopp, M. Erbe, T. F. Sattel, S. Biederer, and T. M. Buzug. Generation of a static magnetic field-free line using two Maxwell coil pairs. *Applied Physics Letters*, 97:092505–1–092505–3, 2010.

[KES⁺11] T. Knopp, M. Erbe, T. F. Sattel, S. Biederer, and T. M. Buzug. A Fourier slice theorem for magnetic particle imaging using a field-free line. *Inverse Problems*, 27(9):095004, 2011.

[KGCZC12] J. Konkle, P. W. Goodwill, O. Carrasco-Zevallos, and S. M. Conolly. Experimental 3D X-Space Magnetic Particle Imaging Using Projection Reconstruction. *Springer Proceedings in Physics 140*, 2012.

[Kno11] T. Knopp. *Effiziente Rekonstruktion und alternative Spulen-topologien für Magnetic-Particle-Imaging.* Springer Vieweg, Berlin/Heidelberg, 2011.

[Kro07] H. Kronmüller. General Micromagnetic Theory in Handbook of Magnetism and Advanced Magnetic Materials. *John Wiley & Sons, New York,* 2007.

[KSBB10] T. Knopp, T. F. Sattel, S. Biederer, and T. M. Buzug. Field-Free Line Formation in a Magnetic Field. *Journal of Physics A: Mathematical and Theoretical,* 43(1):9pp, 2010.

[KSNG99] L. B. Kiss, J. Söderlund, G. A. Niklasson, and C. G. Granqvist. New approach to the origin of lognormal size distributions of nanoparticles. *Nanotechnology,* 10:25–28, 1999.

[Lau73] P. C. Lauterbur. Image Formation by Induced Local Interactions: Examples Employing Nuclear Magnetic Resonance. *Nature,* 242:190–191, doi:10.1038/242190a0, 1973.

[LBBE+10] K. Lüdtke-Buzug, S. Biederer, M. Erbe, T. Knopp, T. F. Sattel, and T. M. Buzug. Superparamagnetic Iron Oxide Nanoparticles for Magnetic Particle Imaging. *Magnetic Nanoparticles: Particle Science, Imaging Technology, and Clinical Applications,* 1:44–50, World Scientific Publishing Company, Singapore, 2010.

[LBF+97] R. Lawaczeck, H. Bauer, T. Frenzel, M. Hasegawa Y. Ito, K. Kito, N. Miwa, H. Tsutsui, H. Vogler, and H.J.Weinmann. Magnetic iron oxide particles coated with carboxydextran for parenteral administration and liver contrasting. *Acta Radiologica,* 38:584–597, 1997.

[Max73] J. C. Maxwell. A Treatise on Electricity and Magnetism. *Clarendon Press, Oxford,* 1873.

[N49] L. Néel. Théorie du traîage magnétique des ferromagnétiques en grains fins avec applications aux terres cuites. *Annales Geophysicae,* 5:99–136, 1949.

[N55] L. Néel. Some theoretical aspects of rock-magnetism. *Advances in Physics,* 4:191–243, doi:10.1080/00018735500101204, 1955.

[R98]   W. C. Röntgen. Über eine neue Art von Strahlen. *Sitzungsberichte der Würzburger Physikalisch-Medizinischen Gesellschaft*, 1898.

[Rad17] J. H. Radon. Über die Bestimmung von Funktionen durch Ihre Integralwerte längs gewisser Mannigfaltigkeiten. *Berichte der Sächsischen Akadamie der Wissenschaft*, 69:262–277, 1917.

[RBK+09] B. Ruhland, K. Baumann, T. Knopp, T. F. Sattel, S. Biederer, K. Luedtke-Buzug, T. M. Buzug, K. Diedrich, and D. Finas. Magnetic Particle Imaging with Superparamagnetic Nanoparticles for sentinel lymph node detection in breast cancer. *Geburtshilfe und Frauenheilkunde*, 69(8):758, doi:10.1055/s-0031-1286434, 2009.

[RGWB10] J. Rahmer, B. Gleich, J. Weizenecker, and J. Borgert. 3D Real-Time Magnetic Particle Imaging of Cerebral Blood Flow in Living Mice. In *Proceedings of the International Society for Magnetic Resonance in Medicine*, volume 18, page 714, Stockholm, 2010.

[RMC03] U. Schwertmann R. M. Cornell. *The Iron Oxides: Structure, Properties, Reactions, Occurences and Uses*. Wiley-VCH, 2nd edition, New York, doi:10.1002/3527602097, 2003.

[Rog11] H. Rogge. *Magnetic Particle Imaging, Diploma thesis*. Institute of Medical Engineering, University of Lübeck, Lübeck, 2011.

[RWGB09] J. Rahmer, J. Weizenecker, B. Gleich, and J. Borgert. Signal encoding in magnetic particle imaging. *BMC Medical Imaging*, 9, 2009.

[RWGB12] J. Rahmer, J. Weizenecker, B. Gleich, and J. Borgert. Analysis of a 3-D system function measured for magnetic particle imaging. *IEEE Transactions on Medical Imaging*, 31(6):1289–99, doi:10.1109/TMI.2012.2188639, 2012.

[SBE+11] T. F. Sattel, S. Biederer, M. Erbe, T. Knopp, and T. M. Buzug. Magnetic Particle Spectrometer with Enlarged Excitation Field Strength. In *Proceedings Biomedizinische Technik / Biomedical Engineering*, volume 56 (Suppl. 1), Freiburg, 2011. Walter de Gruyter, Berlin/New York, doi:10.1515/BMT.2011.275.

[Sch10] H. Schomberg. Magnetic particle imaging: Model and reconstruction. In *Proceedings of the IEEE International Symposium on Biomedical Imaging*, pages 992–995, doi:10.1109/ISBI.2010.5490155, 2010.

[SEB⁺12] T. F. Sattel, M. Erbe, S. Biederer, T. Knopp, D. Finas, K. Diedrich, K. Lüdtke-Buzug, J. Borgert, and T. M. Buzug. Single-sided magnetic particle imaging device for the sentinel lymph node biopsy scenario. In *Proceedings of the SPIE Symposium on Medical Imaging: Biomedical Applications in Molecular, Structural, and Functional Imaging, 83170S*, San Diego, doi:10.1117/12.912733, 2012.

[SGK⁺rg] I. Schmale, B. Gleich, J. Kanzenbach, J. Rahmer, J. Schmidt, J. Weizenecker, and J. Borgert. An Introduction to the Hardware of Magnetic Particle Imaging. In *Proceedings of the World Congress on Medical Physics and Biomedical Engineering, IFMBE Proceedings*, number 25/2, pages 450–453, Munich, 2009. Springer Verlang, Berlin/Heidelberg.

[Sha] L. Shank. AE Techron 7548/7796 Operator's Manual, 2011, `http://aetechron.com/pdf/7548-7796OperatorManual.pdf` (retrieval date: 23.09.2013).

[SHEB13] T. F. Sattel, S. E. Heinitz, M. Erbe, and T. M. Buzug. Experimental Setup of Receive Coils with Transaxial Sensitivity Profiles. In *Proceedings of the International Workshop on Magnetic Particle Imaging*, Berkeley, doi:10.1109/IWMPI.2013.6528355, 2013.

[SKB⁺09] T. F. Sattel, T. Knopp, S. Biederer, B. Gleich, J. Weizenecker, J. Borgert, and T. M. Buzug. Single-Sided Device for Magnetic Particle Imaging. *Journal of Physics D: Applied Physics*, 42(1):1–5, 2009.

[SKB⁺10] T. F. Sattel, T. Knopp, S. Biederer, M. Erbe, K. Lüdtke-Buzug, and T. M. Buzug. Resolution Distribution in Single-Sided Magnetic Particle Imaging. *Magnetic Nanoparticles: Particle Science, Imaging Technology, and Clinical Applications*, 1:106–112, World Scientific Publishing Company, Singapore, 2010.

[SS94] V. I. Stepanov and M. I. Shliomis. *Relaxation Phenomena in Condensed Matter, Chapter: Theory of the Dynamic Syscep-*

tibility of Magnetic Fluids. John Wiley & Sons, New York, doi:10.1002/9780470141465.fmatter, 1994.

[WBG07] J. Weizenecker, J. Borgert, and B. Gleich. A simulation study on the resolution and sensitivity of magnetic particle imaging. *Physics in Medicine and Biology*, 52:6363–6374, 2007.

[Web12] M. Weber. *Power Loss Optimized Field Free Line Generation for Magnetic Particle Imaging.* Master's thesis, Institute of Medical Engineering, University of Lübeck, Lübeck, 2012.

[WEB$^+$54] M. Weber, M. Erbe, K. Bente, T. F. Sattel, and T. M. Buzug. Power Loss Optimized Field Free Line Generation for Magnetic Particle Imaging. In *Proceedings of the International Workshop on Magnetic Particle Imaging*, Berkeley, 2013, doi:10.1109/IWMPI.2013.6528354.

[WGB08] J. Weizenecker, B. Gleich, and J. Borgert. Magnetic particle imaging using a field free line. *Journal of Physics D: Applied Physics*, 41(10):3pp, 2008.

[WGR$^+$09] J. Weizenecker, B. Gleich, J. Rahmer, H. Dahnke, and J. Borgert. Three-dimensional real-time in vivo magnetic particle imaging. *Physics in Medicine and Biology*, 54(5):L1–L10, 2009.

Aktuelle Forschung Medizintechnik

Herausgeber:

Prof. Dr. Thorsten M. Buzug

Institut für Medizintechnik, Universität zu Lübeck

Themen
Werke aus folgenden Themengebieten werden gerne in die Reihe aufgenommen: Biomedizinische Mikro- und Nanosysteme, Elektromedizin, biomedizinische Mess- und Sensortechnik, Monitoring, Lasertechnik, Robotik, minimalinvasive Chirurgie, integrierte OP-Systeme, bildgebende Verfahren, digitale Bildverarbeitung und Visualisierung, Kommunikations- und Informationssysteme, Telemedizin, eHealth und wissensbasierte Systeme, Biosignalverarbeitung, Modellierung und Simulation, Biomechanik, aktive und passive Implantate, Tissue Engineering, Neuroprothetik, Dosimetrie, Strahlenschutz, Strahlentherapie.

Autorinnen und Autoren
Autoren der Reihe sind in der Regel junge Promovierte und Habilitierte, die exzellente Abschlussarbeiten verfasst haben.

Leserschaft
Die Reihe wendet sich einerseits an Studierende, Promovenden und Habilitanden aus den Bereichen Medizintechnik, Medizinische Ingenieurwissenschaft, Medizinische Physik, Medizinische Informatik oder ähnlicher Richtungen. Andererseits stellt die Reihe aktuelle Arbeiten aus einem sich schnell entwickelnden Feld dar, so dass auch Wissenschaftlerinnen und Wissenschaftler sowie Entwicklerinnen und Entwickler an Universitäten, in außeruniversitären Forschungseinrichtungen und der Industrie von den ausgewählten Arbeiten in innovativen Gebieten der Medizintechnik profitieren werden.

Begutachtungsprozess
Die Qualitätssicherung erfolgt in drei Schritten. Zunächst werden nur Arbeiten angenommen die mindestens magna cum laude bewertet sind. Im zweiten Schritt wird ein Mitglied des Editorial Boards die Annahme oder Ablehnung des Werkes empfehlen. Im letzten Schritt wird der Reihenherausgeber über die Annahme oder Ablehnung entscheiden sowie Änderungen in der Druckfassung empfehlen. Die Koordination übernimmt der Reihenherausgeber.

Kontakt
Prof. Dr. Thorsten M. Buzug
Institut für Medizintechnik
Universität zu Lübeck
Ratzeburger Allee 160
23538 Lübeck, Germany

Tel.: +49 (0) 451 / 500-5400
Fax: +49 (0) 451 / 500-5403
E-Mail: buzug@imt.uni-luebeck.de
Web: http://www.imt.uni-luebeck.de

Stand: Januar 2014. Änderungen vorbehalten.
Erhältlich im Buchhandel oder beim Verlag.

Abraham-Lincoln-Straße 46
D-65189 Wiesbaden
Tel. +49 (0)6221. 345 - 4301
www.springer-vieweg.de